Schriften der Mathematisch-naturwissenschaftlichen Klasse
der Heidelberger Akademie der Wissenschaften
Nr. 15 (2004)

Peter Roquette

The
Brauer-Hasse-Noether
Theorem
in Historical Perspective

 Springer

Prof. Dr. Dr. h.c. Peter Roquette
Mathematisches Institut der Universität Heidelberg
Im Neuenheimer Feld 288
69120 Heidelberg, Germany
roquette@uni-hd.de

Library of Congress Control Number: 2004111361

ISBN 3-540-23005-X Springer Berlin Heidelberg New York

Springer is a part of Springer Science+Business Media
springeronline.com
© Springer-Verlag Berlin Heidelberg 2005
Printed in Germany

Cover design: Erich Kirchner, Heidelberg
Typeset in L^ATEX by the author
and edited by PublicationService Gisela Koch, Wiesenbach, using a modified
Springer L^ATEX macro-package.

Printed on acid-free paper 08/3150hs – 5 4 3 2 1 0

Contents

1

Introduction

The legacy of Helmut Hasse, consisting of letters, manuscripts and other papers, is kept at the *Handschriftenabteilung* of the University Library at Göttingen. Hasse had an extensive correspondence; he liked to exchange mathematical ideas, results and methods freely with his colleagues. There are more than 8000 documents preserved. Although not all of them are of equal mathematical interest, searching through this treasure can help us to assess the development of Number Theory through the 1920's and 1930's. Unfortunately, most of the correspondence is preserved on one side only, i.e., the letters *sent to Hasse* are available whereas many of the letters which had been *sent from him*, often handwritten, seem to be lost. So we have to interpolate, as far as possible, from the replies to Hasse and from other contexts, in order to find out what he had written in his outgoing letters.[1]

The present article is largely based on the letters and other documents which I have found concerning the

Brauer-Hasse-Noether Theorem

in the theory of algebras; this covers the years around 1931. Besides the documents from the Hasse and the Brauer legacy in Göttingen, I shall also use some letters from Emmy Noether to Richard Brauer which are preserved at the Bryn Mawr College Library (Pennsylvania, USA).

We should be aware that the Brauer-Hasse-Noether Theorem, although to be rated as a highlight, does not constitute the summit and end point of a development. We have to regard it as a step, important but not final, in a development which leads to the view of class field theory as we see it today. By concentrating on the Brauer-Hasse-Noether Theorem we get only what may be called a snapshot within the great edifice of class field theory.

A snapshot is not a panoramic view. Accordingly, the reader might miss several aspects which also could throw some light on the position of the Brauer-Hasse-Noether theorem, its sources and its consequences, not only within algebraic number theory but also in other mathematical disciplines. It would have

---

[1] An exception is the correspondence between Hasse and Richard Brauer. Thanks to Prof. Fred Brauer, the letters from Hasse to Richard Brauer are now available in Göttingen too.

been impossible to include all these into this paper. Thus I have decided to present it as it is now, being aware of its shortcomings with respect to the range of topics treated, as well as the time span taken into consideration.

A preliminary version of this article had been written in connection with my lecture at the conference March 22–24, 2001 in Stuttgart which was dedicated to the memory of Richard Brauer on the occasion of his centenary. For Brauer, the cooperation with Noether and Hasse in this project constituted an unforgettable, exciting experience. Let us cite from a letter he wrote many years later, on March 3, 1961, to Helmut Hasse: [2]

> *. . . It is now 35 years since you introduced me to class field theory. It belongs to my most delightful memories that I too was able, in cooperation with you and Emmy Noether, to give some little contribution, and I shall never forget the excitement of those days when the paper took shape.*

The available documents indicate that a similar feeling of excitement was present also in the minds of the other actors in this play. Besides Hasse and Noether we have to mention Artin and also Albert in this connection. Other names will appear in due course.

As to A. Adrian Albert, he also had an independent share in the proof of the main theorem, so much that perhaps it would be justified to name it the

Albert-Brauer-Hasse-Noether Theorem.

But since the old name, i.e., without Albert, has become standard in the literature I have abstained from introducing a new name. In mathematics we are used to the fact that the names of results do not always reflect the full story of the historical development. In Section 8, I will describe the role of Albert in the proof of the Brauer-Hasse-Noether theorem, based on the relevant part of the correspondence of Albert with Hasse.

Acknowledgement: Preliminary versions had been in my homepage for some time. I would like to express my thanks to all who sent me their comments each of which I have carefully examined and taken into consideration. Moreover, I wish to thank Falko Lorenz and Keith Conrad for their thorough reading of the last version, their corrections and valuable comments. Last but not least I would like to express my gratitude to Mrs. Nancy Albert, daughter of A. A. Albert, for letting me share her recollections of her father. This was particularly helpful to me while preparing Section 8.

---

[2] The letter is written in German. Here and in the following, whenever we cite a German text from a letter or from a paper then we use our own free translation.

30. Beweis eines Hauptsatzes
in der Theorie der Algebren [52]

gemeinsam mit R. Brauer und E. Noether

Journal für die reine und angewandte Mathematik
167 (1932), 399 – 404

Endlich ist es unseren vereinten Bemühungen gelungen, die Richtigkeit des folgenden Satzes zu beweisen, der für die Strukturtheorie der Algebren über algebraischen Zahlkörpern sowie auch darüber hinaus von grundlegender Bedeutung ist:

Hauptsatz. *Jede normale Divisionsalgebra über einem algebraischen Zahlkörper ist zyklisch (oder, wie man auch sagt, vom Dicksonschen Typus).*

Es ist uns eine besondere Freude, dieses Ergebnis, als einen im wesentlichen der p-adischen Methode zu dankenden Erfolg, Herrn Kurt Hensel, dem Begründer dieser Methode, zu seinem 70. Geburtstag vorzulegen.

Unser Beweis besteht in drei Reduktionen, von denen jeder von uns eine beigesteuert hat [2]).

1. Die erste Reduktion gab H. Hasse auf Grund der von ihm kürzlich entwickelten Theorie der zyklischen Algebren über algebraischen Zahlkörpern [3]).

Reduktion 1. *Der Hauptsatz ist bewiesen, wenn gezeigt ist:*

I. *Jede überall zerfallende Algebra über Ω ist $\sim \Omega$.*

---

[1]) Die Abfassung dieser Note übernahm H. Hasse.

[2]) Diese werden in der Reihenfolge ihrer Entstehung wiedergegeben, die der systematischen Reihenfolge entgegengesetzt ist.

[3]) H. Hasse, Theorie der zyklischen Algebren über einem algebraischen Zahlkörper, Gött. Nachr. 1931. — Eine ausführliche Darstellung der Beweise, in der insbesondere auch die von E. Noether in einer Vorlesung entwickelte, dort wie hier grundlegende Theorie der Zerfällungskörper und verschränkten Produkte entwickelt wird, erscheint demnächst in den Trans. Amer. Math. Soc. (Theory of cyclic algebras over an algebraic number field). Die letztere Arbeit wird im folgenden mit H zitiert. H, 1—6 machen — bis auf den Unterschied in der Sprache — die erstere Note aus.

The Main Theorem: Cyclic Algebras

On December 29, 1931 Kurt Hensel, the mathematician who had discovered p-adic numbers, celebrated his 70th birthday. On this occasion a special volume of Crelle's Journal was dedicated to him since he was the chief editor of Crelle's Journal at that time, and had been for almost 30 years. The dedication volume contains the paper [BrHaNo:1932], authored jointly by Richard Brauer, Helmut Hasse and Emmy Noether, with the title:

Proof of a Main Theorem in the theory of algebras.[3]

The paper starts with the following sentence:

At last our joint endeavours have finally been successful, to prove the following theorem which is of fundamental importance for the structure theory of algebras over number fields, and also beyond . . .

The theorem in question, which has become known as the Brauer-Hasse-Noether Theorem, reads as follows:

Main Theorem [4] *Every central division algebra over a number field is cyclic (or, as it is also said, of Dickson type).*

In this connection, all algebras are assumed to be finite dimensional over a field. An algebra A over a field K is called "central" if K equals the center of A. Actually, in the original Brauer-Hasse-Noether paper [BrHaNo:1932] the word "normal" was used instead of "central"; this had gradually come into use at that time, following the terminology of American authors, see e.g., [Alb:1930].[5] Today the more intuitive "central" is standard.

---

[3] *"Beweis eines Hauptsatzes in der Theorie der Algebren."*

[4] Falko Lorenz [Lor:2004] has criticized the terminology "Main Theorem". Indeed, what today is seen as a "Main Theorem" may in the future be looked at just as a useful lemma. So we should try to invent another name for this theorem, perhaps "Cyclicity Theorem". But for the purpose of the present article, let us keep the authors' terminology and refer to it as the "Main Theorem" (in capitals).

[5] It seems that in 1931 the terminology "normal" was not yet generally accepted. For, when Hasse had sent Noether the manuscript of their joint paper asking for her comments, she suggested that for "German readers" Hasse should explain the notion of "normal". (Letter of November 12, 1931.) Hasse followed her suggestion and inserted an explanation.

Cyclic algebras are defined as follows. Let $L|K$ be a cyclic field extension, of degree n, and let σ denote a generator of its Galois group G. Given any a in the multiplicative group $K^\times$, consider the K-algebra generated by L and some element u with the *defining relations*:

$$u^n = a, \qquad xu = ux^\sigma \quad \text{(for } x \in L\text{)}.$$

This is a central simple algebra of dimension n^2 over K and is denoted by $(L|K, \sigma, a)$. The field L is a maximal commutative subalgebra of $(L|K, \sigma, a)$. This construction of cyclic algebras had been given by Dickson; therefore they were also called "of Dickson type". [6]

Thus the Main Theorem asserts that every central division algebra A over a number field K is isomorphic to $(L|K, \sigma, a)$ for a suitable cyclic extension $L|K$ with generating automorphism σ, and suitable $a \in K^\times$; equivalently, A contains a maximal commutative subfield L which is a cyclic field extension of K.

When Artin heard of the proof of the Main Theorem he wrote to Hasse: [7]

... You cannot imagine how ever so pleased I was about the proof, finally successful, for the cyclic systems. This is the greatest advance in number theory of the last years. My heartfelt congratulations for your proof. ...

Now, given the bare statement of the Main Theorem, Artin's enthusiastic exclamation sounds somewhat exaggerated. At first glance the theorem appears as a rather special result. The description of central simple algebras may have been of importance, but would it qualify for the "greatest advance in number theory in the last years"? It seems that Artin had in mind not only the Main Theorem itself, but also its proof, involving the so-called Local-Global Principle and its many consequences, in particular in class field theory.

The authors themselves, in the first sentence of their joint paper, tell us that they see the importance of the Main Theorem in the following two directions:

1. **Structure of division algebras.** The Main Theorem allows a complete classification of division algebras over a number field by means of what today are called *Hasse invariants*; thereby the structure of the *Brauer group* of an algebraic number field is determined. (This was elaborated in Hasse's subsequent paper [Has:1933] which was dedicated to Emmy Noether on the occasion of her 50th birthday on March 23, 1932.) The splitting fields of a division algebra can be explicitly described by their local behavior; this is important

---

[6] Dickson himself [Dic:1927] called these "algebras of type D". Albert [Alb:1930] gives 1905 as the year when Dickson had discovered this construction. – Dickson did not yet use the notation $(L|K, \sigma, a)$ which seems to have been introduced by Hasse.

[7] This letter from Artin to Hasse is not dated but we have reason to believe that it was written around November 11, 1931.

for the representation theory of groups. (This had been the main motivation for Richard Brauer in this project.)

2. **Beyond the theory of algebras.** The Main Theorem opens new vistas into one of the most exciting areas of algebraic number theory at the time, namely the understanding of *class field theory* – its foundation, its structure and its generalization – by means of the structure of algebras. (This had been suggested for some time by Emmy Noether. It was also Artin's viewpoint when he praised the Main Theorem in the letter we have cited above.)

We will discuss these two viewpoints in more detail in the course of this article.

3

The Paper: Dedication to Hensel

But let us first have a brief look at the dates involved. The *Hensel Festband* of Crelle's Journal carries the publication date of January 6, 1932. The first copy was finished and presented to Hensel already on December 29, 1931, his birthday.[8] The Brauer-Hasse-Noether paper carries the date of receipt of November 11, 1931. Thus the paper was processed and printed within less than two months. This is a remarkably short time for processing and printing, including two times proofreading by the authors. It seems that the authors submitted their paper in the last minute, just in time to be included into the Hensel dedication volume. Why did the authors not submit it earlier? After all, Hasse himself was one of the editors of Crelle's Journal and so he was informed well in advance about the plans for the Hensel dedication volume.

The answer to our question is that the authors did not find their result earlier. For we can determine almost precisely the day when the proof of the Main Theorem had been completed. There is a postcard from Emmy Noether to Hasse dated November 10, 1931 which starts with the following words:

This is beautiful! And completely unexpected to me, notwithstanding that the last argument, due to Brauer, is quite trivial (Every prime number dividing the index is also a divisor of the exponent.) . . .

This is a response to a postcard from Hasse telling her that he had found the last step in the proof of the Main Theorem, by means of an argument which Hasse had learned from Brauer. The theorem of Brauer which she cites in parentheses had been proved in [Bra:1929b]. Of course she does not mean that Brauer's theorem is trivial, but that the *application* of Brauer's theorem in the present situation seems trivial to her. Actually, we shall see in Section 4.2 that this theorem of Brauer is not really needed but only his Sylow argument which he had used in [Bra:1929b].

Only two days earlier, on November 8, 1931, Noether had sent a long letter[9] in which she congratulated Hasse for his recent proof that at least every *abelian*

---

[8] We know this because Hasse mentioned it in his laudation which he read to Hensel on the birthday reception. See [Has:1932].

[9] The letter has 4 pages. This must be considered as "long" by the standard of Emmy Noether who often scribbled her messages on postcards, using up every conceivable free space on the card.

central simple algebra A is cyclic. Here, a central simple algebra $A|K$ is called "abelian" if it admits a splitting field which is an abelian field extension of K.

But Noether did not only congratulate. In addition, she showed Hasse how to obtain a simplification (which she called "trivialization") of his proof, and at the same time to generalize his result from "abelian" to "solvable" algebras by means of an easy induction argument. Moreover, she gave some ideas how it may be possible to approach the general, non-solvable case. These latter ideas were quite different from the final solution which consisted in applying Brauer's Sylow argument; this explains her surprise which she shows in her postcard of November 10.

In those times, postal mail went quite fast. Between Marburg (where Hasse lived) and Göttingen (Noether's place) ordinary mail was delivered the next day after dispatch, sometimes even on the same day.[10]

Thus it appears that on November 9, Hasse had received Noether's earlier letter of November 8. While studying her proofs for the solvable case he remembered an earlier letter of Brauer, where a Sylow argument was used to reduce the general case to the case of a p-group which, after all, is solvable. Putting things together Hasse saw the solution. Brauer's letter had been written some days earlier, on October 29.[11]

Immediately Hasse informed Emmy Noether about his finding, and so it was possible that she received his message on November 10 and could send her reply postcard on the same day.

Accordingly we may conclude that

November 9, 1931

is to be very likely the birthday of the Brauer-Hasse-Noether Main Theorem, i.e., the day when the last step in the proof had been found.

The same day Hasse informed Richard Brauer too. Just two days earlier, on November 7, Hasse had sent a long 10 page letter to Brauer, explaining to him in every detail his ideas for attacking the problem. He used Brauer's Sylow argument but then he said:

> I have to admit that here I am at the end of my skills and I put all my hope on yours. As you see, there are factor systems involved which belong to non-galois splitting fields . . .

Since Brauer had introduced and investigated factor systems for non-galois splitting fields [Bra:1926], [Bra:1928], it appears quite natural that Hasse turned

---

[10] Mail was delivered two times a day: once in the morning and a second time in the afternoon.

[11] Actually, Brauer in his letter did not have the Main Theorem in mind but the related question whether the index of an algebra equals its exponent, over an algebraic number field as a base.

to him for the solution of the problem. But two days later Hasse could send a postcard with the following text:

Dear Mr. Brauer! Just now I receive a letter from Emmy which takes care of the whole question, and such that it will not be necessary to know the structure of the factor systems... It is possible to get a proof by stepwise reduction to steps of prime degree. I had clumsily tried to start with the field below, i.e., with group on top, but it is better to do it the other way... I had gone to many troubles but did not find the simple idea of Emmy.

And Hasse continued to describe Emmy's idea, all on the same postcard.

Brauer lived in Königsberg which was somewhat more distant to Marburg than Göttingen; thus the postcard to him may have needed one day longer than that to Noether. [12] In fact, Brauer's reply to Hasse is dated November 11, one day later than Noether's reply. He wrote:

Many thanks for your detailed letter, and for your postcard which I just received. It is very nice that the problem of cyclicity is now solved! Just today I had meant to write you and to inform you in detail about Emmy's method; but I have to admit that I feared to make a silly mistake because I had the feeling that the thing was too simple. I just wanted to ask you about it, but now this is unnecessary. By the way, right from the beginning it was clear to me that with your reduction, the essential work had been done already...

This shows that Brauer was directly involved in finding the proof. When he speaks of "Emmy's method" he refers to Hasse's postcard where Hasse had mentioned the reduction step sent to him by Emmy Noether. But in fact, Brauer had found the same method independently and he too had realized that, if combined with his Sylow argument, this method would give the solution. At the same time we see his modesty, which made him claim that Hasse had done the essential work already. Two days later when Hasse had sent him the completed manuscript, he wrote:

Today in the morning I received your paper; I am quite surprised that my really small remark has caused you to publish this particularly beautiful paper jointly under my name...

Well, Brauer's contribution was not confined to a "small remark". On the contrary, Hasse's arguments relied heavily and substantially on Brauer's general results about division algebras and their splitting fields.

We have seen that the "birthday" of the Main Theorem had been November 9, but we have also seen that the manuscript was received by the editors on November 11. We conclude that Hasse had completed the manuscript in at most

---

[12] From Marburg to Göttingen there are about 140 km, whereas from Marburg to Königsberg we have counted about 975 km.

two days. Actually, it must have been within one day because on November 11 already, Emmy Noether had received from him the completed manuscript and wrote another letter to Hasse with her comments. This haste is explained by the fact that the deadline for contributions to the Hensel dedication volume had passed long ago (it was September 1, 1931) and Hensel's birthday was approaching at the end of the year already, when the volume had to be presented to him. And Hasse was eager to put this paper, which he considered important, into this dedication volume; Kurt Hensel had been his respected academic teacher and now was his paternal friend (*"väterlicher Freund"*). In the introduction of the Brauer-Hasse-Noether paper we read:

> *It gives us particular pleasure to be able to dedicate this result, being essentially due to the p-adic method, to the founder of this method, Mr. Kurt Hensel, on the occasion of his 70th birthday.*

Emmy Noether commented on this dedication text in her letter of November 12 to Hasse as follows:

> *Of course I agree with the bow to Hensel. My methods are working- and conceptual methods* [13] *and therefore have anonymously penetrated everywhere.*

The second sentence in this comment has become famous in the Noether literature. It puts into evidence that she was very sure about the power and success of "her methods" which she describes quite to the point. But why did she write this sentence just here, while discussing the dedication text for Hensel? The answer which suggests itself is that, on the one hand, Noether wishes to express to Hasse that, after all, "her methods" (as distinguished from Hensel's p-adic methods) were equally responsible for their success. On the other hand she does not care whether this is publicly acknowledged or not.

In the present context "her methods" means two things: First, she insists that the classical representation theory be done within the framework of the abstract theory of algebras (or hypercomplex systems in her terminology), instead of matrix groups and semi-groups as Schur had started it. Second, she strongly proposes that the non-commutative theory of algebras should be used for a better understanding of commutative algebraic number theory, in particular class field theory.

Perhaps we may add a third aspect of "her methods": the power to transmit her ideas and concepts to the people around her. In this way she had decisively influenced Richard Brauer's and Helmut Hasse's way of thinking: Brauer investigated division algebras and Hasse did non-commutative arithmetic.

The great hurry in which the Brauer-Hasse-Noether paper had to be written may also account for the somewhat unconventional presentation. For, Hasse says in a footnote that the material is presented

---

[13] *"Arbeits- und Auffassungsmethoden".*

. . . in the order of the discovery, which is the reverse of the systematic order. . . .

This footnote was inserted on the insistence of Noether. For, in still another letter written 3 days later, on November 14, 1931, she had expressed her *dislike* of the presentation as given by Hasse. She wrote that in this presentation the proof is difficult to understand, and that she would have insisted on a more systematic arrangement except that the time was too short. Therefore Hasse should at least insert a footnote to the effect as mentioned above. And Hasse did so. He wished the paper to be included into the Hensel volume, hence there was no time to rewrite the manuscript.

Three months later Hasse seized an opportunity to become reconciled with Emmy Noether by dedicating a new paper [Has:1933] to her, on the occasion of her 50th birthday on March 23, 1932. There he deals with the same subject but written more systematically. Those three months had seen a rapid development of the subject; in particular Hasse was now able to give a proof of Artin's Reciprocity Law of class field theory which was based almost entirely on the theory of the Brauer group over a number field. Thereby he could fulfill a *desideratum* of Emmy Noether who already one year earlier had asked him to give a hypercomplex foundation of the reciprocity law. In the preface to that paper Hasse "bows" to Emmy Noether as an invaluable source of inspiration.

Section II.6 of that paper [Has:1933] contains a new presentation of the Main Theorem. Hasse starts this section by admitting that in the earlier joint paper [BrHaNo:1932] the proof had been presented in a somewhat awkward manner, according to the order of its discovery. Now, he says, he will give the proof (which is the same proof after all) in a more systematic way. Clearly, this is to be viewed as a response to Noether's criticism in her letter of November 14.

By the way, three days after her birthday Emmy Noether replied to this present: "I was terribly delighted!. . . " (*Ich habe mich schrecklich gefreut!. . .*). There follow two pages of detailed comments to Hasse's paper, showing that she had studied it already in detail.

The Local-Global Principle

Let K be an algebraic number field of finite degree. For every prime $\mathfrak{p}$ of K, finite or infinite, let $K_{\mathfrak{p}}$ denote the $\mathfrak{p}$-adic completion of K. For an algebra A over K we put $A_{\mathfrak{p}} = A \otimes_K K_{\mathfrak{p}}$, the completion (also called localization) of A at $\mathfrak{p}$. An important step in the proof of the Main Theorem is the celebrated

Local-Global Principle for algebras *Let K be a number field and $A|K$ be a central simple algebra. If $A_{\mathfrak{p}}|K_{\mathfrak{p}}$ splits for every $\mathfrak{p}$ then $A|K$ splits.*

Here, "splitting" of $A|K$ means that A is a full matrix algebra over K. Note that the Local-Global Principle is formulated for simple algebras, not only for division algebras as the Main Theorem had been. Quite generally, it is more convenient to work with simple algebras and, accordingly, formulate and prove the Main Theorem for simple algebras instead of division algebras only. By Wedderburn's theorem, every simple algebra $A|K$ is isomorphic to a full matrix ring over a division algebra $D|K$, and D is uniquely determined by A up to isomorphisms. Two central simple algebras over K are called "similar" if their corresponding division algebras are isomorphic.

We shall discuss in Section 5 how the Local-Global Principle was used in the proof of the Main Theorem. In the present section we review the long way which finally led to the conception and the proof of the Local-Global Principle.

4.1 The Norm Theorem

First, consider a cyclic algebra $A = (L|K, \sigma, a)$ as explained in Section 2. Such an algebra splits if and only if a is a norm from the cyclic extension $L|K$. Accordingly, the Local-Global Principle for cyclic algebras can be reformulated as follows:

Hilbert-Furtwängler-Hasse Norm Theorem *Let $L|K$ be a cyclic extension of number fields, and let $0 \neq a \in K$. If a is a norm in the completion $L_{\mathfrak{p}}|K_{\mathfrak{p}}$ for every $\mathfrak{p}$ then a is a norm in $L|K$.*[14]

---

[14] In accordance with the definition of $A_{\mathfrak{p}}$ one would define $L_{\mathfrak{p}} = L \otimes_K K_{\mathfrak{p}}$. In general this is not a field but the direct sum of fields $L_{\mathfrak{P}}$ where $\mathfrak{P}$ ranges over the primes of

This theorem does not refer to algebras, it concerns algebraic number fields only. Now, in the case when the degree n of $L \mid K$ is a prime number, the Norm Theorem was known for a long time already, in the context of the reciprocity law of class field theory. It had been included in Hasse's class field report, Part II [Has:1930] where Hasse mentioned that it had first been proved by Furtwängler in [Fur:1902] and subsequent papers. For quadratic fields ($n = 2$) the Norm Theorem had been given by Hilbert in his *Zahlbericht* [Hil:1897]. In March 1931 Hasse succeeded to generalize this statement to arbitrary cyclic extensions $L \mid K$ of number fields, not necessarily of prime degree; see section 7. He published this in [Has:1931a], April 1931.

Now, the Main Theorem tells us that *every* central simple algebra over a number field is cyclic, so we could conclude that the Local-Global Principle holds generally, for every central simple algebra over a number field. However, in order to *prove* the Main Theorem, Hasse needed *first* to prove the Local-Global Principle generally, regardless of whether the given algebra is already known to be cyclic or not. Hence there arose quite naturally the problem how to reduce the general case of the Local-Global Principle to the case when the algebra is cyclic.

4.2 The Reductions

In the Brauer-Hasse-Noether paper [BrNo:1927] this reduction is done in two steps:

(2) *Reduction to the case when A has a solvable splitting field.*

(3) *Further reduction (by induction) to the case when A has a cyclic splitting field.*

Here and in the following we use the same enumeration of these "reductions" which is used in the Brauer-Hasse-Noether paper. In that enumeration "reduction (1)" would be the contribution of Hasse; this we will discuss in Section 5 and the following sections.

The reduction (2) is due to Brauer who, in his letter to Hasse of October 29, 1931, had provided a Sylow argument for this purpose. Reduction (3) had been provided by Noether in her letter to Hasse of November 8, 1931.

Brauer had developed the theory of division algebras and matrix algebras in a series of several papers in the foregoing years, starting from his 1927 *Habilitationsschrift* at the University of Königsberg [Bra:1928]. His main interest was in the theory of group representations, following the ideas of his academic

---

L dividing $\mathfrak{p}$. If $L \mid K$ is a Galois extension (in particular if it is cyclic) then all these fields $L_{\mathfrak{P}}$ are isomorphic over $K_{\mathfrak{p}}$, and a is a norm from $L_{\mathfrak{p}}$ if and only if it is a norm from $L_{\mathfrak{P}}$ for some and hence all $\mathfrak{P}$. Usually, one chooses one prime $\mathfrak{P}\mid\mathfrak{p}$ and writes $L_{\mathfrak{p}}$ for $L_{\mathfrak{P}}$ (thus forgetting the former, systematic notation for $L_{\mathfrak{p}}$). Let us do this here too.

teacher I. Schur. It was Emmy Noether who gradually had convinced him that the representation theory of groups could and should be profitably discussed within the framework of algebras. In Brauer's papers, in particular in [Bra:1929b], one can find the following theorems. Brauer had reported on these theorems in September 1928 at the annual meeting of the DMV (*Deutsche Mathematiker Vereinigung*) in Hamburg; see [Bra:1929a]. Although in that report no proofs are given, we can recommend consulting it since Brauer's theorems are very clearly stated there.

Brauer's theorems. (i) *The similarity classes of central simple algebras over a field K form a group with multiplication well-defined by the tensor product $A \otimes_K B$ of two algebras.*[15]

Today this group is called the "Brauer group" of K and denoted by $Br(K)$. The name "Brauer group" was given by Hasse in [BrHaNo:1932]. The split algebras belong to the neutral element of the Brauer group.

(ii) *Every central simple algebra A over K has finite order in $Br(K)$.*

This order is called the "exponent" of A. This terminology had been chosen by Brauer because, he said, in the context of the theory of algebras the word "order" is used for another concept.[16]

(iii) *The exponent of A divides the index m of A.*

The index m of A is defined as follows: Let D be the division algebra similar to A. The dimension of D over its center is a square m^2, and this m is the index by definition, also called "Schur index".

(iv) *Every prime number dividing the index of A also divides its exponent.*

Brauer had used these theorems (i)–(iv) in order to show:

(v) *Every division algebra A of index m can be decomposed as the tensor product of division algebras A_i of prime power index $p_i^{\nu_i}$, according to the prime power decomposition $m = \prod_i p_i^{\nu_i}$ of the index.*

In Hasse's first draft of the joint manuscript which he had sent to Emmy Noether, these theorems were used. Although we do not know this first draft, we can conclude this from the following: First, in Noether's reply postcard of November 10 (which we have cited in Section 3) Brauer's theorem (iv) is mentioned. Secondly, in a letter of Hasse to Brauer dated November 11, Hasse reports that Noether

---

[15] At the time of Brauer-Hasse-Noether, the tensor product was called "direct product" and denoted by $A \times B$. – Brauer considered only perfect base fields K; it was Emmy Noether who in [Noe:1929] was able to wave the hypothesis that K is perfect.

[16] I am indebted to Falko Lorenz who pointed out to me that this theorem (ii) is contained in Schur's paper [Schu:1919] already, as well as theorem (iii) if m is interpreted suitably. See [Lor:1998].

had finally thrown out the reduction (v) to prime power index, because that was superfluous. And so Hasse continues:

> For that reason I did not find a suitable occasion to cite your paper of Mathematische Zeitschrift [Bra: 1929b]. Almost nothing from there is needed, except the most simple facts on splitting fields.

These "most simple facts" which are used in the final proof are the following:

> The degree of every splitting field of A over K is divisible by the index m of A, and there exist splitting fields of degree m.

Using this, the reduction steps (2) and (3) are quite easy if combined with the functorial properties of the Brauer group. Let us briefly present the arguments. Our presentation is the same as Noether had proposed it in her letter to Hasse of November 10, 1931, and which Hasse then used in his Noether dedication paper [Has: 1933].

If $K \subset L$ then we use the notation $A_L = A \otimes_K L$. If we regard A and A_L in their respective Brauer groups $Br(K)$ and $Br(L)$ then the map $A \mapsto A_L$ defines a canonical group homomorphism $Br(K) \to Br(L)$.

Let K be a number field and $A|K$ a central simple algebra which splits everywhere locally. The claim is that A splits. Suppose A does not split and let $m > 1$ be the index of A. Let p be a prime number dividing m. Consider a Galois splitting field $L|K$ of A, so that A_L splits; then p divides $[L : K]$. Let G be the Galois group of $L|K$. Consider a corresponding p-Sylow subgroup of G and let $L_0 \subset L$ denote its fixed field. Since the p-Sylow group is solvable there exists a chain of fields

$$L_0 \subset L_1 \subset \cdots \subset L_{s-1} \subset L_s = L$$

such that each $L_i|L_{i-1}$ is cyclic of degree p ($1 \leq i \leq s$). Since A splits everywhere locally, so does every A_{L_i}. Now, $A_{L_{s-1}}$ has $L_s = L$ as a cyclic splitting field. Hence the Norm Theorem implies that $A_{L_{s-1}}$ splits. Therefore $A_{L_{s-2}}$ has L_{s-1} as a cyclic splitting field, hence again, $A_{L_{s-2}}$ splits. And so on by induction. Finally we conclude that A_{L_0} splits. Thus A admits the splitting field L_0 whose degree $[L_0 : K]$ is relatively prime to p. But $[L_0 : K]$ is divisible by the index m which contains p as a prime divisor. Contradiction.

Since each $L_i|L_{i-1}$ is of degree p, it is evident that Hasse's Norm Theorem has to be used only in the case of cyclic fields *of prime degree p*, i.e., *the original Hilbert-Furtwängler Theorem is sufficient*. Hasse's generalization to arbitrary cyclic fields is not needed and is a consequence of Noether's induction argument. This had been immediately observed by Noether (letter of November 8, 1931), and she had asked Hasse to mention it in their joint paper (which he did). At that time this observation indeed could be considered a simplification. But half a year later, in [Has: 1933], Hasse remarked that this would not make a difference any more because in the meantime new proofs of the Norm Theorem had been found by Chevalley and Herbrand, and those proofs work equally well for arbitrary cyclic extensions (using the so-called Herbrand's Lemma) regardless of whether the degree is prime or not. [17]

From today's viewpoint the above proof of the Local-Global Principle looks rather trivial, once the Hilbert-Furtwängler Norm Theorem is accepted. In particular if the arguments are given in the language of cohomology, as it is usually done nowadays, we see that only the very basic properties of the cohomological restriction map are used. This seems to justify Brauer's words, cited above, that *"right from the beginning it was clear to me* [Brauer] *that with your* [Hasse's] *reduction, the essential work had been done already"*. But these words are valid only if, firstly, Brauer's fundamental theorems are accepted and, secondly, there had already developed a certain routine for using those theorems for particular problems. While the first was certainly the case within the circle around Brauer, the second was not yet. Otherwise, the simple proof above could well have been given much earlier.

We should not underestimate the conceptual difficulty which people had at that time working with algebras and their splitting fields, and the notions of index and exponent of algebras. There was no established routine to work with the functorial properties of Brauer groups. Based on the cited work of Brauer and, in parallel, on the monumental work of Emmy Noether [Noe:1929] such routine came gradually into being.

4.3 Factor Systems

The idea for a proof like the above required, in the first place, some insight into the relevant structures, in particular *the interpretation of the Norm Theorem as a splitting theorem for cyclic algebras*, before trying to generalize it from the cyclic to the general case. In fact, Hasse originally did not do this step. In his class field report Part II [Has:1930] he had conjectured that the Norm Theorem holds for arbitrary abelian extension of number fields. But in [Has:1931a] he had to admit that for non-cyclic extensions the Norm Theorem fails to hold.

It was Emmy Noether who then suggested to Hasse that the generalization of the Norm Theorem would require considering algebras instead of norms, the latter representing split cyclic algebras. This is evidenced by the following excerpt from her letter of November 12, 1931. In that letter she wished to have some further changes in the manuscript of the joint Brauer-Hasse-Noether paper, for which Hasse had composed the draft. She wrote:

> *... Similarly, I would like to be mentioned too on page 4, in the 4-th paragraph from below, or maybe the "H. Hasse" should be replaced by "we"* [18]. *For,*

---

[17] The Herbrand-Chevalley proof was included in Hasse's Marburg lectures 1932 on class field theory. See [Has:1933a], Satz (113).

[18] In the printed version, this is the last paragraph of Section 4. There indeed we find the word "we" as Noether requested.

I have mentioned to you already in the spring on our Hanstein-walk [19] that the version with factor systems is the correct *generalization, after you had told me the refutation of the norm conjecture in the abelian case. Perhaps you had not yet fully grasped it at the time, and later you have come to the same conclusion by yourself. Strictly speaking I had mentioned this to you already in Nidden...* [20]

We observe that Noether talks about *factor systems* and not about algebras. Factor systems are used to construct algebras. Given any finite separable field extension $L|K$ let $Br(L|K)$ denote the kernel of the map $Br(K) \to Br(L)$, consisting of those central simple algebras over K (modulo similarity) which are split by L. Brauer had shown that $Br(L|K)$ is isomorphic to the group of what he called *factor systems* (modulo equivalence). A factor system consists of certain elements in the Galois closure of $L|K$, and it can be used to construct a central simple algebra $A|K$ such that the elements of the factor system appear as factors in the defining relations of a suitable basis of the algebra. Brauer's invention of factor systems was essential for the proof of his theorems.

It is true that the appearance of factor systems had been observed earlier already by Schur and also by Dickson. But it was Brauer who defined and used them systematically to construct algebras, thereby writing down explicitly the so-called associativity relations.

We will not give here the explicit definition of factor systems in the sense of Brauer. For, today one mostly uses in this context the simplified form which Noether has given to Brauer's factor systems. Noether considered Galois splitting fields $L|K$ only. Let $G = G(L|K)$ denote its Galois group. Consider the K-algebra A which is generated by L and by elements u_σ ($\sigma \in G$) with the defining relations

$$u_\sigma u_\tau = u_{\sigma\tau} a_{\sigma,\tau}, \qquad x u_\sigma = u_\sigma x^\sigma \quad \text{(for } x \in L)$$

where $\sigma, \tau \in G$ and $a_{\sigma,\tau} \in K^\times$. It is required that the factors $a_{\sigma,\tau}$ satisfy the following relations which are called *associativity relations*:

$$a_{\sigma,\tau\varrho} \cdot a_{\tau,\varrho} = a_{\sigma\tau,\varrho} \cdot a_{\sigma,\tau}^\varrho .$$

Sometimes they are also called *Noether equations*. The algebra A thus defined is a central simple algebra over K which has L as a maximal commutative subfield.

---

[19] Hanstein is a hillside near Göttingen. It appears that in the spring of 1931, on one of the many visits of Hasse to Göttingen, they had made a joint excursion to the Hanstein.

[20] Nidden at that time was a small fisherman's village in East Prussia, located on a peninsula (*Kurische Nehrung*) in the Baltic sea and famed for its extended white sand dunes. In September 1930, Hasse and Noether both attended the annual meeting of the DMV at Königsberg in East Prussia, and after the meeting they visited Nidden together.

A is called the "crossed product" of L with its Galois group G, and with factor system $a = (a_{\sigma,\tau})$.[21] Notation: $A = (L|K, a)$. Every central simple algebra over K which admits L as a splitting field can be represented, up to similarity, as a crossed product in this sense. If G is cyclic then (by appropriate choice of the u_{σ}) we obtain the cyclic algebras in this way.

This theory of factor systems was developed by Emmy Noether in her Göttingen lecture 1929/30. But Noether herself never published her theory. Deuring took notes of that lecture, and these were distributed among interested people; Brauer as well as Hasse had obtained a copy of those notes. (The Deuring notes are now included in Noether's Collected Papers.) The first publication of Noether's theory of crossed products was given, with Noether's permission, in Hasse's American paper [Has:1932a] where a whole chapter is devoted to it. The theory was also included in the book "*Algebren*" by Deuring [Deu:1934].

A factor system $a_{\sigma,\tau}$ is said to split if there exist elements $c_{\sigma} \in L^{\times}$ such that

$$a_{\sigma,\tau} = \frac{c_{\sigma}^{\tau} c_{\tau}}{c_{\sigma\tau}}.$$

The split factor systems are those whose crossed product algebra $(L|K, a)$ splits. Today we view the group of factor systems modulo split ones as the second cohomology group of the Galois group G of $L|K$ in the multiplicative G-module $L^{\times}$. The notation is $H^2(G, L^{\times})$ or better $H^2(L|K)$.

Thus the Brauer-Noether theory of crossed products yields an isomorphism

$$Br(L|K) \approx H^2(L|K)$$

which has turned out to be basic for Brauer's theory.

In mathematics we often observe that a particular object can be looked at from different points of view. A change of viewpoint may sometimes generate new analogies, thereby we may see that certain methods had been successfully applied in similar looking situations and we try to use those methods, suitably modified, to deal also with the problem at hand. This indeed can lead the way to new discoveries. But sometimes it can also hamper the way because the chosen analogies create difficulties which are inessential to the original problem.

We can observe such a situation in Hasse's first attempts to deal with the Local-Global Principle for algebras. Instead of dealing with algebras directly he considered, following Noether's suggestion, factor systems. Given a factor

---

[21] The German terminology is "*verschränktes Produkt*". The English term "crossed product" had been used by Hasse in his American paper [Has:1932a]. When Noether read this she wrote to Hasse: "*Are the 'crossed products' your English invention? This word is good.*" We do not know whether Hasse himself invented this terminology, or perhaps it was H. T. Engstrom, the American mathematician who helped Hasse to translate his manuscript from German into English. In any case, in the English language the terminology "crossed product" has been in use since then.

system in $H^2(L|K)$ which splits locally everywhere, he tried to transform it in such a way that its global splitting is evident. This then reduces to the solution of certain diophantine equations in L, e.g., the Noether equations, under the hypothesis that those equations can be solved locally everywhere. Now it is well known, and it was of course known to Hasse, that the local solution of diophantine equations does *not* in general imply their global solution. But several years earlier Hasse had already proved one instance of a Local-Global Principle for certain diophantine equations, namely *quadratic equations*. In case of the rational field $\mathbb{Q}$ as base field this had been the subject of Hasse's dissertation (Ph. D. thesis) in 1921, and in subsequent papers [Has:1924a], [Has:1924b] he solved the same problem for an arbitrary number field K as base field.[22]

Accordingly, Hasse tried first to invoke the analogy to the theory of *quadratic forms* in order to approach the Local-Global Principle *for algebras*. However, it turned out that this created difficulties which only later were seen not to be inherent to the problem.

We are able to follow Hasse's ideas for these first attempts (which later were discarded as unnecessary) since Hasse had written to some of his friends explaining these ideas, obviously in the hope that someone would be able to supply the final clue. One of those letters, the one to Brauer dated July 27, 1931, is preserved. Hasse writes:

> ... *I would like to write to you about the only question which is still open, the question whether all central* [23] *simple algebras are cyclic. For I believe that this question is now ripe and I would like to present to you the line of attack which I have in mind.*

(In this connection Hasse means algebras over an algebraic number field as base field, although he does not explicitly mention this.)

---

[22] With this result Hasse had solved, at least partially, one of the famous Hilbert problems. The 13th Hilbert problem calls for solving "*a given quadratic equation with algebraic numerical coefficients in any number of variables by integral or fractional numbers belonging to the algebraic realm of rationality determined by the coefficients.*" Hilbert's wording admits two interpretations. One of them is to regard the phrase "integer or fractional numbers" as denoting arbitrary numbers of the number field in question. In this interpretation Hasse could be said to have solved the problem completely. The other interpretation is that Hilbert actually meant two different problems: The first is to solve the quadratic equation in integers of the field, and the second is to admit solutions with arbitrary numbers of the field. In this interpretation, which would generalize Minkowski's work from the rationals to arbitrary number fields, Hasse would have solved only the second of the two problems. The first problem (solution in integers) has been studied by Siegel and others.

[23] Hasse writes "normal" instead of "central". For the convenience of the reader we will replace "normal" in this context by the modern "central", here and also in other citations which follow.

Hasse continues that, following his "line of attack", he is trying to use his Local-Global Principle for quadratic forms. Let w_i be a basis of the given central simple algebra A over K. The trace matrix $\text{tr}(w_i w_k)$ defines a quadratic form. If A splits locally everywhere then for every prime $\mathfrak{p}$ there exists a basis transformation which transforms the given basis into a system of matrix units, and this defines a certain transformation of the quadratic form. The Local-Global Principle for quadratic forms then yields a certain basis transformation over the global field K. Hasse asks whether it is possible to deduce the splitting of A from the special structure of this transformed trace form. In other words, one has to construct from it a system of matrix units over K. But Hasse does not yet know how to do this, not even whether it is possible at all. He writes to Brauer:

I would like to put this problem into your hands for examination from this viewpoint.

Several days later, on August 3, 1931 Brauer replied that at present he is not able to say anything about Hasse's problem, and that first he has to study it in detail. But relying on Hasse's own creative power he adds:

I hope to be able to understand all these things at the time when you will have filled the gap yourself.

Hasse had sent similar letters to Artin and Noether. These letters are not preserved but we know the respective answers. Artin, returning from a vacation in the mountains wrote on August 24, 1931:

... Meanwhile you will certainly have proved the theorem on division algebras. I am looking forward to it.

Noether wrote on the same day:

Naturally, I too cannot answer your question – I believe one should leave such things alone until one meets them again from another point of view ...

But she adds some remarks about the work of Levitzky (her Ph. D. student) who provided some methods to construct bases of split algebras.

These answers do not sound as if they had been very helpful to Hasse. But he did not give up so easily. After a while he managed to prove the Local-Global Principle for those algebras A which admit an *abelian* splitting field $L|K$. We do not know this proof but from Noether's reaction we can infer that indeed Hasse had explicitly constructed, by induction, a split factor system for the algebra. We have already mentioned in Section 3 (p.10) Noether's reaction to Hasse's letter; the Noether letter was dated November 8, 1931 and gave a simplification and generalization of Hasse's result to algebras which admit a *solvable* splitting field, not necessary abelian.

From then on things began to develop rapidly as we have explained in Section 3, and one day later the proof of the Local-Global Principle was complete.

As a side remark we mention that Hasse in his letter to Brauer of November 16, 1931 states that when Noether's postcard arrived on November 9 he had "essentially been through" with his complicated proof. But, as we have seen, he immediately threw away his complicated proof in favor of Noether's "trivialization".

From the Local-Global Principle
to the Main Theorem

Sometimes the Local-Global Principle is considered the most important result of the Brauer-Hasse-Noether paper while the Main Theorem is rated as just one of the many consequences of it. But the authors themselves present the Main Theorem as their key result. We now discuss the step from the Local-Global Principle to the Main Theorem. This is the "reduction (1)" in the count of the Brauer-Hasse-Noether paper, and it is due to Hasse.

5.1 The Splitting Criterion

Let A be a central simple algebra over a number field K. Then $[A : K]$ is a square; let $[A : K] = n^2$ with $n \in \mathbb{N}$. It is known that n is a multiple of the index m of A. In order to show that A is cyclic one has to construct a cyclic splitting field $L|K$ of A of degree $[L : K] = n$. To this end one needs a criterion for a finite extension field L of K to be a splitting field of A.

According to the Local-Global Principle the problem can be shifted to the local completions, namely:

A is split by L if and only if each $A_{\mathfrak{p}}$ is split by $L_{\mathfrak{P}}$ for $\mathfrak{P}|\mathfrak{p}$.

In the local case, there is a simple criterion for splitting fields:

Local Splitting Criterion *$A_{\mathfrak{p}}$ is split by $L_{\mathfrak{P}}$ if and only if the degree $[L_{\mathfrak{P}} : K_{\mathfrak{p}}]$ is divisible by the index $m_{\mathfrak{p}}$ of $A_{\mathfrak{p}}$.*

Thus the Local-Global Principle yields:

Global Splitting Criterion *A is split by L if and only if for each prime $\mathfrak{p}$ of K and each $\mathfrak{P}$ dividing $\mathfrak{p}$ the local degree $[L_{\mathfrak{P}} : K_{\mathfrak{p}}]$ is divisible by the local index $m_{\mathfrak{p}}$ of $A_{\mathfrak{p}}$.*

If $L|K$ is a Galois extension then for all primes $\mathfrak{P}$ dividing $\mathfrak{p}$ the completions $L_{\mathfrak{P}}$ coincide; they may be denoted by $L_{\mathfrak{p}}$ according to the notation explained in footnote 14.

The local criterion was essentially contained already in Hasse's seminal Annalen paper [Has:1931] on the structure of division algebras over local fields. But the criterion was not explicitly stated there. Therefore Hasse in their joint paper [BrHaNo:1932] gave a detailed proof of the criterion, based on the main results of [Has:1931]. But again it was not explicitly stated; instead, the statement and proof was embedded in the proof of the global criterion which was "Satz 3" in the joint paper.

So the local criterion, although it is one of the basic foundations on which the Main Theorem rests, remained somewhat hidden in the Hasse-Brauer-Noether paper – another sign that the preparation of the manuscript was done in great haste. It was so well hidden that even five months later Emmy Noether was not aware that its proof was contained in the paper of which she was a co-author after all. In her letter of April 27, 1932 she wrote, referring to a recent paper of Köthe:[24]

> ... In fact, Köthe with his theorem on invariants shows directly, that in the p-adic case the degree condition is also sufficient for splitting fields ...

With "invariants" are meant what today are called the "Hasse invariants" of central simple algebras over a local field $K_\mathfrak{p}$ (see Section 6.1, p. 40 below). Köthe's theorem in [Koe:1933] describes the effect of a base field extension to these invariants. If the base field $K_\mathfrak{p}$ is extended to a finite extension $L_\mathfrak{p}$ then, according to Köthe's theorem, the Hasse invariant of the extended algebra $A_{L_\mathfrak{p}}$ is obtained from the Hasse invariant of $A_\mathfrak{p}$ by multiplication with the field degree $[L_\mathfrak{p} : K_\mathfrak{p}]$. This implies the local splitting criterion.

One week later Noether admitted that she had overlooked Hasse's proof in the joint paper [BrHaNo:1932]. Obviously responding to a reproach of Hasse she wrote:

> ... When I got Köthe's paper it occurred to me that now this old question was settled. In yours I had overlooked it; or, what is more likely, I thought about my old proof and had only skimmed through yours. [25]

Now let us return to the Global Splitting Criterion. Its degree conditions are non-trivial only for the primes $\mathfrak{p}$ for which the local index $m_\mathfrak{p} > 1$. For a given central simple algebra there are only finitely many such primes. This is by no means trivial; it had been proved by Hasse in [Has:1931] where he showed that the reduced discriminant ("*Grundideal*") of a maximal order of A contains $\mathfrak{p}$ to the exponent $m_\mathfrak{p} - 1$. We conclude that the existence of a cyclic splitting field $L|K$ of degree n for A is equivalent to the following general

---

[24] Gottfried Köthe (1905–1989) was a young post-doc who in 1928/1929 came to Göttingen to study mainly with Emmy Noether and van der Waerden. Later he switched to functional analysis under the influence of Toeplitz.

[25] Since Noether had wished to inform Hasse about Köthe's results it seems that she did not know (or not remember) that Köthe's paper [Koe:1933] was written largely under the influence and the guidance of Hasse. This is expressed by Köthe in a footnote to his paper which appeared in the *Mathematische Annalen* right after Hasse's [Has:1933].

Existence Theorem *Let K be an algebraic number field and S a finite set of primes of K. For each $\mathfrak{p} \in S$ let there be given a number $m_\mathfrak{p} \in \mathbb{N}$.* [26] *Moreover, let $n \in \mathbb{N}$ be a common multiple of the $m_\mathfrak{p}$'s. Then there exists a cyclic field extension $L|K$ of degree n such that for each $\mathfrak{p} \in S$ the local degree $[L_\mathfrak{p} : K_\mathfrak{p}]$ is a multiple of $m_\mathfrak{p}$.*

This then settles the Main Theorem.

The Existence Theorem as such does not refer to algebras. It belongs to algebraic number theory. We shall discuss the theorem and its history in the next sections.

5.2 An Unproven Theorem

A proof of the Existence Theorem had been outlined in a letter of Hasse to Albert written in April 1931. This is reported in the paper [AlHa:1932]. But the proof is not given and not even outlined in [AlHa:1932]. Hasse did not publish a proof of his existence theorem, not in the joint paper [BrHaNo:1932] and not elsewhere. Why not? After all, the existence theorem is an indispensable link in the chain of arguments leading to the proof of the Main Theorem. Without it, the proof of the Main Theorem would be incomplete. Now, in a footnote in [AlHa:1932] we read:

> *The existence theorem is a generalization of those in Hasse's papers* [Has:1926a], [Has:1926b] *and a complete proof will be published elsewhere.*

This remark gives us a clue why Hasse may have hesitated to publish his proof prematurely. He regarded his existence theorem as an integral part of number theory and was looking for the most general such theorem, independently of its application to the proof of the Main Theorem. We shall see that Grunwald, a Ph. D. student of Hasse, provided such a very general theorem. This then leads to the Grunwald-Wang story.

The story begins with a reference which Hasse had inserted in the Brauer-Hasse-Noether paper [BrHaNo:1932] for a possible proof of the Existence Theorem. This reference reads: "[vgl.d.Anm.zu H,17Bb]". This somewhat cryptical reference can be decoded as: "compare the footnote in the paper H, section 17, Proof of (17.5) part B, subsection (b)." The code "H" refers to Hasse's American paper [Has:1932a] on cyclic algebras. That paper had not yet appeared at the time when he wrote down the manuscript for the Brauer-Hasse-Noether paper, hence he could not give a page number. We have checked that the page number is 205. But there, in the said footnote of [Has:1932a] it is merely stated: "*The*

---

[26] For infinite primes the usual resrictions should be observed: If $\mathfrak{p}$ is real then $m_\mathfrak{p} = 1$ or 2; if $\mathfrak{p}$ is complex then $m_\mathfrak{p} = 1$. This guarantees that in any case $m_\mathfrak{p}$ is the index of some central simple algebra over $K_\mathfrak{p}$.

existence of such a field will be proved in another place." [27] This does not sound very helpful to the reader.

Let us check the next paper [Has:1933] of Hasse. This is the one which he had dedicated to Emmy Noether and in which, among other topics, he repeats the proof of the Main Theorem more systematically. There he says at the corresponding spot on page 519:

> *Such a sufficiently strong existence theorem has been proved recently by Engström [1]. Alternatively, it is possible to deduce such a theorem, probably in its greatest possible generality, from the thesis of Grunwald [1] which has recently appeared; see Grunwald [2].*

Checking the bibliography of [Has:1933] we find under "Engstrom [1]" the entry: "Publication in an American journal in preparation." However we were not able to find, either in an American journal or elsewhere, any publication of H. T. Engstrom where this or a similar theorem is proved.

Howard T. Engstrom was a young American postdoc from Yale who had stayed in Göttingen for the academic year 1931. Through Emmy Noether he got in contact with Hasse. He had helped translating Hasse's American paper [Has:1933] into English. Emmy Noether wrote about him in a letter of June 2, 1931:

> *Engstrom was quite satisfied with your English, apart from the rearrangements; hopefully you will be satisfied with his existence theorems too! He is really very enthusiastic about everything which he had learned in Germany. I am sending Engstrom's manuscript to Deuring who for a long time is waiting impatiently for it ...*

It appears that Hasse had proposed to Engstrom to write up the proof of the Existence Theorem according to his (Hasse's) outline, and that Deuring was to check Engstrom's manuscript.

But Engstrom did not complete his manuscript before he returned to Yale. We have found a letter from Engstrom to Hasse, dated February 27, 1932 from Yale, where he apologizes that he has not finished the manuscript on existence theorems as yet. He concludes:

> *Your outline indicates to me that you have expended considerable thought on the matter, and that it would really require not much effort on your part to write it up for publication. If this is the case please don't hesitate to do so....*

We get the impression that Deuring had found a flaw in Engstrom's manuscript and that Hasse had given Engstrom some hints how to overcome the difficulty.

---

[27] The footnote continues to announce that this existence theorem will be another one in a series of former existence theorems proved by Hasse – same remark as we had already seen above in the paper [AlHa:1932].

But at Yale Engstrom was absorbed by different duties and, hence, returned the subject to Hasse.

It remains to check Grunwald, the second reference which was mentioned by Hasse in [Has:1933].

5.3 The Grunwald-Wang Story

Grunwald had been a Ph. D. student with Hasse at the University of Halle, and he had followed Hasse to Marburg in 1930. The reference "Grunwald [1]" in Hasse's paper refers to his thesis [Gru:1932] which appeared 1932 in the *Mathematische Annalen*. The subject of the thesis belongs to the fundamentals of algebraic number theory; from today's viewpoint it can be viewed as a first attempt to understand the role of Hecke's *Größencharaktere* in class field theory. Grunwald's thesis does not contain the Existence Theorem, but Hasse discovered that Grunwald's methods could be used to obtain a proof of the theorem. From the correspondence between Grunwald and Hasse (which is preserved) we can infer that Hasse had proposed to Grunwald to extract from his thesis a proof of the Existence Theorem and publish it in a separate paper.

And Grunwald did so. The reference "Grunwald [2]" in Hasse's [Has:1933] refers to Grunwald's paper, at that time still "forthcoming", which appeared 1933 in Crelle's Journal [Gru:1933]. There Grunwald proved a general existence theorem which became known as "Grunwald's theorem". This theorem is much stronger than Hasse's Existence Theorem:

> **Grunwald's theorem** *Let K be an algebraic number field and S a finite set of primes of K. For each $\mathfrak{p} \in S$ let there be given a cyclic field extension $L_\mathfrak{p}|K_\mathfrak{p}$. Moreover, let $n \in \mathbb{N}$ be a common multiple of the degrees $[L_\mathfrak{p} : K_\mathfrak{p}]$. Then there exists a cyclic field extension $L|K$ of degree n such that for each $\mathfrak{p} \in S$ its completion coincides with the given fields $L_\mathfrak{p}$.*

Whereas Hasse needed only the fact that the local degrees $[L_\mathfrak{p} : K_\mathfrak{p}]$ should be multiples of the given numbers $m_\mathfrak{p}$ (for $\mathfrak{p} \in S$), Grunwald's theorem claims that even the local fields $L_\mathfrak{p}$ themselves can be prescribed as cyclic extensions of degree $m_\mathfrak{p}$ of $K_\mathfrak{p}$ (for the finitely many primes $\mathfrak{p} \in S$). This was a beautiful and strong theorem, and clearly it settled the question.[28]

The proof of Grunwald's theorem used class field theory and was considered to be quite difficult. In 1942 a simplified proof was given by Whaples [Wha:1942]; it also used class field theory but no analytic number theory which had still been necessary at the time of Grunwald.

---

[28] Wilhelm Grunwald (1909–1989) did not continue to work in Mathematics but decided to become a science librarian. He finally advanced to the position of director of the renowned Göttingen University Library but he always preserved his love for Mathematics, in particular Number Theory. He kept contact with Hasse throughout his life.

In the year 1948 Artin, who was at Princeton University at that time, conducted a seminar on class field theory. One of the seminar talks was devoted to Whaples' new proof of Grunwald's theorem. Here is what happened in the seminar, told by one of the participants, John Tate[29]:

> *I had just switched from physics to math, and tried to follow it* [the seminar] *as best I could. Wang also attended that seminar. In the spring of 1948, Bill Mills, one of the students Artin had brought with him from Indiana, talked on "Grunwald's Theorem" in the seminar. Some days later I was with Artin in his office when Wang appeared. He said he had a counterexample to a lemma which had been used in the proof. An hour or two later, he produced a counterexample to the Theorem itself... Of course he* [Artin] *was astonished, as were all of us students, that a famous theorem with two published proofs, one of which we had all heard in the seminar without our noticing anything, could be wrong. But it was a good lesson!*

The error was not contained in Grunwald's paper [Gru:1933] itself but in Grunwald's thesis [Gru:1932] from which the author cited a lemma. That lemma referred to a prime number p but the author did not see that the prime $p = 2$ needed special care when compared to the odd primes $p > 2$.

The fact that there was an error in Grunwald's (as well as in Whaples') theorem caused a great stir among the people concerned. Would this mean that the Main Theorem of Brauer-Hasse-Noether was wrong too?

Fortunately, the situation was not that serious. In "most" cases, Grunwald's theorem holds, and exceptions can only occur if n is divisible by 8. Also, Hasse's Existence Theorem is much weaker than Grunwald's and it turned out that this weaker theorem holds in any case, including the cases where the full Grunwald theorem collapses. This was established by Hasse in [Has:1950] immediately after Wang's counter example became known. Moreover, Hasse carefully examined the situation and gave a correct version of Grunwald's theorem, explicitly analyzing the exceptions. Independently of Hasse, Wang too gave a correct formulation in his Ph. D. thesis [Wan:1950]. Since then the corrected theorem is called the "Grunwald-Wang Theorem".[30]

And the Main Theorem of Brauer-Hasse-Noether was saved.

---

[29] In a personal letter to the author.

[30] Shianghao Wang (1915–1993) received his Ph. D. at Princeton University in 1949 and afterwards returned to China. He published two more papers connected with the Grunwald-Wang theorem but later turned to Computer Science, in particular control theory. He was professor and chairman of the Math. Dept. at Jiling Universiy since 1952; vice president 1980/81. He became a member of the Academia Sinica. It is said that "Wang was a versatile person. He was good at chess, bridge, novels, Chinese opera." – I am indebted to Professors Eng Tjioe Tan and Ming-chang Kang for information about the vita of Shianghao Wang.

5.4 Remarks

Let us add some more remarks to this story. Even before the error in Grunwald's theorem was found, Hasse seemed not to be satisfied with its proof. In his opinion the proof as given by Grunwald, which used a lot of class field theory, was not adequate as a basis for such a fundamental result like the Main Theorem. Therefore he thought of ways to avoid the application of Grunwald's theorem in the context of the Main Theorem, if possible.

5.4.1 The Weak Existence Theorem

The Existence Theorem can be weakened by removing the requirement for $L|K$ to have a fixed degree n. In this weak form one is looking for a cyclic extension $L|K$ whose degree is not specified, with the only condition that its local degrees $[L_\mathfrak{p} : K_\mathfrak{p}]$ should be divisible by the given numbers $m_\mathfrak{p}$ (for $\mathfrak{p} \in S$). Already in 1932 Hasse had given a relatively elementary proof of this weak form [Has:1933]. This does not yield the full Main Theorem but only its weak form that every central simple algebra A is *similar* to a cyclic algebra (which *a priori* does not necessarily imply A itself to be cyclic). Quite often this weak form of the Main Theorem turns out to be sufficient in the applications, and so Hasse's proof in [Has:1933] provides a simplified access to those applications, without using the complicated class field proof via the Grunwald-Wang theorem.

Moreover, in order to satisfy this weak form of Hasse's Existence Theorem, it turns out that the required cyclic extension $L|K$ can be constructed as a *cyclotomic extension* of K, i.e., a subfield of $K(\sqrt[\ell]{1})$ for suitable ℓ (which may even be chosen as a prime number).[31] This fact became important when Hasse gave a proof of Artin's Reciprocity Law within the framework of the theory of algebras, as he did in [Has:1933]. See Section 6.2.

5.4.2 Group Representations

In the joint Brauer-Hasse-Noether paper there is a section titled "Applications" and it is attributed to Hasse. One of those applications concerns representations of finite groups:

> *Every absolutely irreducible matrix representation of a finite group G can be realized (up to equivalence) in the field of n^h roots of unity where n is the order of G and h is sufficiently large.*

---

[31] The proof needed a lemma on prime numbers satisfying certain congruence relations, of similar kind as Artin had to use for his general reciprocity law. Simplified proofs of this lemma were later given by Hasse himself, by Chevalley, Iyanaga and finally in greater generality by van der Waerden [vdW:1934].

It is understood that representations are to be meant over a field of character-istic 0. [32] Two representations over the same field are equivalent if they can be transformed into each other by a non-singular matrix; in "modern" terms: if they determine isomorphic G-modules.

This theorem had been conjectured by I. Schur [Schu:1906] but with $h = 1$. In view of this Hasse did not say that his theorem is a proof of Schur's conjecture; instead he says that this theorem constitutes a "support" of Schur's conjecture.

In order to prove the theorem, let $K = \mathbb{Q}(\sqrt[n]{1})$. Hasse considered the group algebra $K[G] = \oplus \sum_i A_i$, decomposed into its simple components.[33] The cen-ter of A_i is K. The assertion of the above theorem is now transformed to say that $K(\sqrt[nh]{1})$ is a splitting field of each A_i, provided h is sufficiently large. Applying the Global Splitting Criterion this means that for each prime $\mathfrak{p}$ of K the $\mathfrak{p}$-local degree of $K(\sqrt[nh]{1})$ over K is a multiple of the local index $m_{i,\mathfrak{p}}$. Only the prime divisors $\mathfrak{p}$ of n are relevant and for those, $m_{i,\mathfrak{p}}$ is seen to divide the group order n. Thus one has to prove the lemma that n divides the local degree $K(\sqrt[nh]{1})$ for each $\mathfrak{p}|n$ and h sufficiently large. This then is easily checked by the known decomposition behavior of primes in cyclotomic fields.[34]

When sending his draft of their joint manuscript to Brauer on November 11, 1931, Hasse wrote:

> I could imagine that you would perhaps have a comment to this theorem or maybe a sharper result. I have considered the question only very roughly. Have I cited I. Schur correctly?

Brauer replied on November 13:

> I was particularly impressed by this theorem; I did not believe that one could apply the methods that directly to this problem. It would be very interesting if one could prove $h = 1$. At present I cannot do this but I will think about it as soon as the commotion of the semester beginning is over – of course, only in case that you have not solved the question in the meantime which I believe is quite probable.

Well, Hasse did not solve the question but it was Brauer who many years later verified the Schur conjecture [Bra:1945]. Two more years later he proved even the stronger result that the field of d-th roots of unity suffices where d is the exponent of the group G [Bra:1947]. See also [Roq:1952] and [BrTa:1955].

This may be an appropriate occasion to cite a letter which Carl Ludwig Siegel wrote to Hasse on December 7, 1931 when he had been informed about

---

[32] At that time the theory of "modular" representations, i.e., over fields of characteristic $p > 0$, had not yet been developed. It had been systematically started by Richard Brauer in the late 1930s.

[33] $\oplus \sum$ is our notation for direct sum.

[34] Actually, the proof in [BrHaNo:1932] of the last mentioned lemma contains an error. Hasse corrects this error in [Has:1949].

the Brauer-Hasse-Noether paper. It seems that in the past Siegel too had tried to prove Schur's conjecture, but without success.

> *Dear Mr. Hasse! ... This is indeed the nicest birthday gift for Hensel, that his p-adic methods have been developed to such triumph. I had not even been able to approach the Schur problem properly... The pessimism which I harbor generally towards the prospects of mathematics has again been shaken ...*

There had been an exchange of letters between Hasse and Siegel before this. When in June 1931 Siegel visited Marburg, Hasse told him about his attempts, unsuccessful at that time, to prove the Local-Global Principle for algebras. On his return to Frankfurt, Siegel wrote a postcard to Hasse with a proof that the discriminant of any division algebra D over a number field is of absolute value > 1; this would have settled the problem at least if D is central over $\mathbb{Q}$. But after examining Siegel's proof Hasse pointed out to him that this proof does not work, which Siegel conceded (*"Many thanks for your exposition of my unsuccessful proof!"*).

5.4.3 Algebras with Pure Maximal Subfields

In 1934 there appeared a paper by Albert with the title "Kummer fields" [Alb:1934]. There, Albert proved the following theorem:

> *A central division algebra D of prime degree p over a field K of characteristic 0 is cyclic if and only if D contains an element $x \notin K$ such that $x^p \in K$.*[35]

Of course, this is trivial if K contains the p-th roots of unity because then $K(x)$ would be a cyclic subfield of D of degree p. If K does not contain the p-th roots of unity then Albert constructs a cyclic field $L|K$ contained in D such that $L(\sqrt[p]{1}) = K(x, \sqrt[p]{1})$; this can be done by the classical methods of Kummer. Albert formulated this theorem over fields of characteristic 0 only but from its proof it was immediately clear that it remains true over all fields of characteristic $\neq p$.[36]

Concerning this theorem, Hasse wrote to Albert in February 1935:

> *Your result seems to me of particular interest. It allows one to eliminate Grunwald's complicated existence theorem in the proof that every central division algebra D of prime degree p over an algebraic number field K is cyclic.*

---

[35] Albert calls $K(x)$ a "pure" extension of K since it is generated by radicals.

[36] In Hasse's mathematical diary, dated "February 1935" we find the following entry: "*Proof of a theorem of Albert, following E. Witt.*" Witt's proof of Albert's theorem is particularly simple, following the style of Emmy Noether, but the essential ingredients are the same as in Albert's proof. It seems that Witt had presented this proof in the Hasse seminar, and that Hasse had noted it in his diary for future reference.

And Hasse proceeds to explain how to derive the Main Theorem for division algebras of degree p from Albert's result. This is easy enough. For, let S denote the set of those primes $\mathfrak{p}$ of K for which the local index $m_\mathfrak{p}$ of D is $\neq 1$. Choose $\pi \in K$ which is a prime element for every finite $\mathfrak{p} \in S$, and $\pi < 0$ for every infinite real $\mathfrak{p} \in S$. Then $K(\sqrt[p]{\pi})$ splits D by the Splitting Theorem, hence $K(\sqrt[p]{\pi})$ is isomorphic to a subfield of D. Applying Albert's theorem it follows that D is cyclic. Hasse continues:

> *We are trying to generalize your theorem to prime power degree. This would eliminate Grunwald's theorem altogether for the proof of the Main Theorem.*

We do not know whether Albert replied to this letter of Hasse. But three years later in [Alb:1938] he showed that Hasse's idea could not be realized. He presented an example of a non-cyclic division algebra of index 4 containing a pure subfield of degree 4. The base field K is the rational function field in three variables over a formally real field.

5.4.4 Exponent = Index

One of the important consequences of the Main Theorem is the fact that over number fields, the exponent of a central simple algebra equals its index. This is a very remarkable theorem. It has interesting consequences in the representation theory of finite groups, and this was the reason why Richard Brauer was particularly interested in it. The theorem does not hold over arbitrary fields since Brauer [Bra:1933] has shown that over function fields of sufficiently many variables, there are division algebras whose exponent e and index m are arbitrarily prescribed, subject only to the conditions which are given in Brauer's theorems which we have cited in Section 4.2 (p. 17). See also [Alb:1932b] where a similar question is studied. [37]

In the Brauer-Hasse-Noether paper the exponent-index theorem is obtained by using the Existence Theorem (see Section 5.1, p. 26). To this end the number n in the Existence Theorem is chosen as the least common multiple of the local indices $m_\mathfrak{p}$ of the $\mathfrak{p}$-components $A_\mathfrak{p}$. Then L splits A by the Splitting Theorem and hence n is a multiple of the index m of A, hence also of its exponent e. On the other hand, from Hasse's local theory [Has:1931] it follows that the local index $m_\mathfrak{p}$ equals the local exponent $e_\mathfrak{p}$ of $A_\mathfrak{p}$. From $A^e \sim 1$ it follows $A_\mathfrak{p}^e \sim 1$ for each $\mathfrak{p}$; therefore e is a multiple of $e_\mathfrak{p}$ (for all $\mathfrak{p}$) and therefore also of n. It follows $e = n = m$.

From the above sequence of arguments it is immediate that in fact it is not necessary to know that $L \,|\, K$ is cyclic. But if cyclicity is not required then it is easy,

---

[37] From the references in those papers it appears that Brauer and Albert did not know the results of each other concerning this question.

by means of the Chinese remainder theorem [38], to construct a field extension $L|K$ of degree n (as above) with the given local degrees $m_{\mathfrak{p}}$ for finitely many primes $\mathfrak{p}$. This was pointed put by Hasse in his letter to Albert of February 1935 which we had already cited above. Hasse wrote:

> *In my Annalen paper* [Has:1933] *I derived theorem* (6.43) *(exponent = index) from Grunwald's existence theorem. In point of fact this deep existence theorem is not necessary for proving index = exponent. For one can carry through the proof with any sort of splitting field L instead of a cyclic L. See my first existence theorem* [Has:1926a].[39]

5.4.5 Grunwald-Wang in the Setting of Valuation Theory

Both proofs of the Grunwald-Wang Theorem, the proof [Wan:1950] by Wang himself and Hasse's proof [Has:1950], use heavy machinery of class field theory. The same is true with Artin's proof in the Artin-Tate lecture notes [ArTa:1968] where there is a whole chapter devoted to the Grunwald-Wang theorem. But the question arises whether the Grunwald-Wang theorem does really belong to class field theory, or perhaps it is valid in a more general setting, for arbitrary fields with valuations. If so then it is to be expected that the proof would become simpler and more adequate. Therefore Hasse [Has:1950] wondered whether it would be possible to give an algebraic proof using Kummer theory instead of class field theory. This is indeed possible and has been shown by the author in collaboration with Falko Lorenz in [LoRo:2003]. See also the literature cited there.

---

[38] Since there are also infinite primes $\mathfrak{p}$ involved, the "Chinese remainder theorem" has to be interpreted such as to include infinite primes too. In other words: this is the theorem of independence of finitely many valuations.

[39] Hasse's first existence theorem is stated and proved for finite primes only, i.e., prime ideals in the base field. It seems that Hasse himself, when he cites his paper in the letter to Albert, regarded the inclusion of infinite primes in his first existence theorem as trivial (which it is).

The Brauer Group and Class Field Theory

The Main Theorem allows us to determine completely the structure of the Brauer group $Br(K)$ of a number field K. As we have reported in Section 2 already, the authors of the Brauer-Hasse-Noether paper regard this as one of the important applications of the Main Theorem.

Let K be a number field and $\mathfrak{p}$ a prime of K. If we associate to every central simple algebra A over K its completion $A_{\mathfrak{p}}$ then we obtain the $\mathfrak{p}$-adic localization map of Brauer groups $Br(K) \to Br(K_{\mathfrak{p}})$. Combining these maps for all primes $\mathfrak{p}$ of K we obtain the universal localization map

$$Br(K) \longrightarrow \oplus \sum_{\mathfrak{p}} Br(K_{\mathfrak{p}})$$

where the sum on the right hand side is understood to be the direct sum. (In this context the Brauer group is written additively.) The Local-Global Principle can be interpreted to say that this localization map is *injective*. Thus $Br(K)$ can be viewed as a subgroup of the direct sum of the local groups $Br(K_{\mathfrak{p}})$. Accordingly, the determination of the structure of $Br(K)$ starts with the determination of the structure of the local components $Br(K_{\mathfrak{p}})$.

6.1 The Local Hasse Invariant

First we consider the case when $\mathfrak{p}$ is a finite prime of K.

The description of $Br(K_{\mathfrak{p}})$ had essentially been done in a former paper by Hasse [Has:1931] in the *Mathematische Annalen*. We have already had occasion to mention this paper in Section 5.1 when we discussed the Local Splitting Criterion. In fact, that criterion is a consequence of the following Local Structure Theorem from [Has:1931].

We denote by $K_{\mathfrak{p}}^{(n)}$ the unramified extension of $K_{\mathfrak{p}}$ of degree n. It is cyclic, and the Galois group is generated by the Frobenius automorphism; let us call it φ. [40] In [Has:1931] we find the following

---

[40] More precisely, we should perhaps write $\varphi^{(n)}$ since it is an automorphism of $K_{\mathfrak{p}}^{(n)}$ depending on n. But let us interpret the symbol φ as the Frobenius automorphism of the maximal unramified extension of $K_{\mathfrak{p}}$; if applied to the elements of $K_{\mathfrak{p}}^{(n)}$ this gives the Frobenius automorphism of $K_{\mathfrak{p}}^{(n)}$. This simplifies the notation somewhat.

Local Structure Theorem (level n)   *Let $A_{\mathfrak{p}}$ be any central simple algebra over $K_{\mathfrak{p}}$, of dimension n^2. Then $A_{\mathfrak{p}}$ contains a maximal commutative subfield isomorphic to $K_{\mathfrak{p}}^{(n)}$. Consequently $A_{\mathfrak{p}}$ is cyclic and admits a representation of the form $A_{\mathfrak{p}} = (K_{\mathfrak{p}}^{(n)}, \varphi, a)$ with $a \in K_{\mathfrak{p}}^{\times}$.*

The remarkable fact is not only that all of those algebras $A_{\mathfrak{p}}$ are cyclic, but that each of them contains the same canonical field extension $K_{\mathfrak{p}}^{(n)}$ as a maximal cyclic subfield. Even more remarkable is how Hasse had derived this. Namely, he applied the classical p-adic methods of Hensel to the non-commutative case. Let us explain this:

The general case is readily reduced to the case when $A_{\mathfrak{p}} = D_{\mathfrak{p}}$ is a *division algebra* of dimension n^2 over $K_{\mathfrak{p}}$. Now, $K_{\mathfrak{p}}$ being a complete field, it carries canonically a valuation which we denote by v. Writing this valuation additively, the axioms for the valuation are

$$v(ab) = v(a) + v(b)$$
$$v(a + b) \geq \min(v(a), v(b)).$$

Now, Hasse's method consisted of extending this valuation to the given division algebra $D_{\mathfrak{p}}$. It turns out that such an extension is uniquely possible; the formula for the extended valuation is

$$v(x) = \frac{1}{n} v(Nx) \qquad\qquad (x \in D_{\mathfrak{p}})$$

where N denotes the reduced norm from $D_{\mathfrak{p}}$ to $K_{\mathfrak{p}}$. This formula and the proof are precisely the same as developed by Hensel for extending valuations to commutative extensions, in particular it uses *Hensel's Lemma*. Now $D_{\mathfrak{p}}$ appears as a *valued skew field* with center $K_{\mathfrak{p}}$. As such it has a *ramification degree e* and a *residue degree f*. But unlike the commutative case it turns out that here, $K_{\mathfrak{p}}$ being the center of $D_{\mathfrak{p}}$, we have always $e = f = n$. Since $f = n$ it follows readily from Hensel's Lemma that $D_{\mathfrak{p}}$ contains the unramified field extension $K_{\mathfrak{p}}^{(n)}$ of degree n, as announced in the theorem.

We have said above that this proof is remarkable. This does not mean that the proof is difficult; in fact, it is straightforward for anyone who is acquainted with Hensel's method of handling valuations. The remarkable thing is that Hasse used *valuations* to investigate *non-commutative* division algebras over local fields. [41] The valuation ring of $D_{\mathfrak{p}}$ consists of all $x \in D_{\mathfrak{p}}$ with $v(x) \geq 0$. It contains a unique maximal ideal, which is a 2-sided ideal, consisting of all x with $v(x) > 0$.

---

[41] After Hasse, the valuation theory of non-commutative structures developed rapidly, not only over number fields but over arbitrary fields. We refer to the impressive report of Wadsworth [Wad:2002] about this development. All this started with Hasse's paper [Has:1931] which is under discussion here.

Remark: Hasse denotes this valuation prime ideal by the letter $\wp$, and this shows up in the title of his paper [Has:1931]. This somewhat strange notation is explained by the fact that, in Hasse's paper, the symbol $\mathfrak{p}$ is used for the canonical prime ideal in the complete field $K_\mathfrak{p}$, and the corresponding capital letter $\mathfrak{P}$ was used to denote the valuation ideal in commutative field extensions. Thus, in order to indicate that in the non-commutative case the situation was somewhat different, Hasse proposed to use a different symbol, and he chose $\wp$ for this purpose. Formerly this symbol, known as the "Weierstrass-p", was used to denote the elliptic function $\wp(z)$ in the Weierstrass normalization. Hasse's notation for the prime ideal of a valued division algebra did not survive, but the Weierstrass notation $\wp(z)$ is still in use today in the theory of elliptic functions.

Emmy Noether, after having read Hasse's manuscript for [Has:1931], recognized immediately its strong potential. Hasse had sent this manuscript to her for publication in the *Mathematische Annalen* for which Noether acted as unofficial editor. [42] On a postcard dated June 25, 1930 she wrote to Hasse: "*Dear Mr. Hasse! I have found your paper on hypercomplex $\mathfrak{p}$-adics very enjoyable.*" [43] And, as it was her custom, she immediately jotted down her comments and proposals for further studies. About the local theory, which concerns us here, she wrote:

> ... *From local class field theory it follows: If $L_\mathfrak{p}$ is cyclic of degree n over a $\mathfrak{p}$-adic base field $K_\mathfrak{p}$ then there exists at least one element $a \neq 0$ such that only a^n becomes a norm of an $L_\mathfrak{p}$-element. Are you able to prove this directly? Then one could derive local class field theory from your skew field results.*

In other words: Noether asks whether the Brauer group $Br(L_\mathfrak{p}|K_\mathfrak{p})$ contains an element of exponent n.

We do not know precisely what Hasse replied to her. But from later correspondence with Noether we can implicitly conclude that he replied something like "I do not know". It took him some time to follow her hint and to realize that he could have said "yes" in view of the Local Structure Theorem above.

Let us briefly indicate the arguments which Hasse could have used. These can be found in Hasse's later papers, the Brauer-Hasse-Noether paper [BrHaNo:1932], the American paper [Has:1932a] and the paper [Has:1932c]

---

[42] "Unofficial" means that her name was not mentioned officially on the title page. But people who knew sent their paper to her if it belonged to Noether's field of interest. As a rule, Noether read the paper and, if she found it suitable, sent it to Blumenthal who, as the managing editor, accepted it. The date of "received by the editors" was set as the date when the paper was received by Emmy Noether. Thus Hasse's paper [Has:1931] carries the date of June 16, 1930.

[43] Emmy Noether used the symbol $\mathfrak{p}$ (in German handwriting) since she did not know how to write $\wp$ as she admitted in a later letter. Can we conclude from this that she never had worked with elliptic functions?

dedicated to Noether's 50th birthday. But in fact the arguments are essentially based on the Local Structure Theorem.

If we associate to each $a \in K_{\mathfrak{p}}^{\times}$ the cyclic algebra $(K_{\mathfrak{p}}^{(n)}, \varphi, a)$ then we obtain a homomorphism from $K_{\mathfrak{p}}^{\times}$ to the Brauer group $Br(K_{\mathfrak{p}}^{(n)}|K_{\mathfrak{p}})$. By the Local Structure Theorem this homomorphism is *surjective*. Its kernel is the group of norms from $K_{\mathfrak{p}}^{(n)}$. But $K_{\mathfrak{p}}^{(n)}$ is *unramified* over $K_{\mathfrak{p}}$ and therefore an element $a \in K_{\mathfrak{p}}^{\times}$ is a norm if and only if its value $v(a) \equiv 0 \bmod n$. [44] Consequently, for any $a \in K_{\mathfrak{p}}^{\times}$ its value $v(a)$ modulo n represents its class in the norm class group, and hence represents the algebra $(K_{\mathfrak{p}}^{(n)}, \varphi, a)$. In particular we see that $Br(K_{\mathfrak{p}}^{(n)}|K_{\mathfrak{p}})$ is isomorphic to $\mathbb{Z}/n$.

By the Local Structure Theorem, $Br(K_{\mathfrak{p}}^{(n)}|K_{\mathfrak{p}})$ contains *all* central simple algebras $A_{\mathfrak{p}}$ of index dividing n. Consequently, if $L_{\mathfrak{p}}$ is an arbitrary extension of $K_{\mathfrak{p}}$ of degree $[L_{\mathfrak{p}} : K_{\mathfrak{p}}] = n$ then the Brauer group

$$Br(L_{\mathfrak{p}}|K_{\mathfrak{p}}) \subset Br(K_{\mathfrak{p}}^{(n)}|K_{\mathfrak{p}}) \approx \mathbb{Z}/n.$$

Hence, Noether's question is answered affirmatively if we have equality here, which is to say that every algebra $A_{\mathfrak{p}}$ in $Br(K_{\mathfrak{p}}^{(n)}|K_{\mathfrak{p}})$ is split by $L_{\mathfrak{p}}$. Since the index of $A_{\mathfrak{p}}$ divides $n = [L_{\mathfrak{p}} : K_{\mathfrak{p}}]$ this follows from the Local Splitting Criterion.

Thus indeed, Hasse could have answered Noether's question with "yes", already in 1930; in fact he did so later. But Noether's conclusion that one could derive local class field theory from this, was too optimistic. Noether's question was concerned with *cyclic* extensions only, but class field theory deals with arbitrary *abelian* extensions. It was only later that Chevalley [Che:1933] showed how to perform the transition from cyclic to arbitrary abelian extensions.

Let us return to the Local Structure Theorem for level n. We have seen above that this implies an isomorphism $Br(K_{\mathfrak{p}}^{(n)}|K_{\mathfrak{p}}) \approx \mathbb{Z}/n$, via the map $(K_{\mathfrak{p}}^{(n)}, \varphi, a) \mapsto v(a) \bmod n$. In other words: the residue class $v(a) \bmod n$ is an invariant of the algebra $A_{\mathfrak{p}} = (K_{\mathfrak{p}}^{(n)}, \varphi, a)$. To obtain an invariant which is independent of n Hasse divided $v(a)$ by n and thus defined what today is called the *Hasse-invariant*:

$$\left(\frac{A_{\mathfrak{p}}}{\mathfrak{p}}\right) :\equiv \frac{v(a)}{n} \bmod \mathbb{Z};$$

this is a certain rational number which is determined modulo integers only. If m is a multiple of n then every $A_{\mathfrak{p}} \in Br(K_{\mathfrak{p}}^{(n)}|K_{\mathfrak{p}})$ can also be viewed to be contained in $Br(K_{\mathfrak{p}}^{(m)}|K_{\mathfrak{p}})$ and it turns out that the Hasse invariant as defined above is the same for m as that for n. If $n \to \infty$ the final version of the Local Structure Theorem emerges:

---

[44] We assume here that the valuation of $K_{\mathfrak{p}}$ is *normalized* such that its value group is $\mathbb{Z}$.

Local Structure Theorem   *Let $\mathfrak{p}$ be a finite prime of K. If we associate to every central simple algebra over $K_{\mathfrak{p}}$ its Hasse invariant then we obtain the canonical group isomorphism*

$$\mathrm{inv}_{\mathfrak{p}} : Br(K_{\mathfrak{p}}) \xrightarrow{\approx} \mathbb{Q}/\mathbb{Z}$$

If $\mathfrak{p}$ is an infinite real prime then $K_{\mathfrak{p}} = \mathbb{R}$ and there is only one non-trivial central division algebra over $\mathbb{R}$, namely the quaternions of Hamilton. If we associate to this the Hasse-invariant $\frac{1}{2}$ then we obtain

$$\mathrm{inv}_{\mathfrak{p}} : Br(K_{\mathfrak{p}}) \xrightarrow{\approx} \begin{cases} \frac{1}{2}\mathbb{Z}/\mathbb{Z} & \text{if } \mathfrak{p} \text{ is real,} \\ 0 & \text{if } \mathfrak{p} \text{ is complex.} \end{cases}$$

6.2 Structure of the Global Brauer Group

Having settled the local structure theorems, Hasse turns now to the global structure, i.e., the structure of $Br(K)$ for a number field K. If A is a central simple algebra over K and $A_{\mathfrak{p}}$ its $\mathfrak{p}$-adic completion then Hasse writes briefly $\left(\frac{A}{\mathfrak{p}}\right)$ instead of $\left(\frac{A_{\mathfrak{p}}}{\mathfrak{p}}\right)$. If we associate to each A its local Hasse invariants $\left(\frac{A}{\mathfrak{p}}\right)$ then we obtain the global invariant map

$$\mathrm{inv} : Br(K) \longrightarrow \oplus \sum_{\mathfrak{p}}{}' \mathbb{Q}/\mathbb{Z}$$

where the prime on the Σ sign of the direct sum should remind the reader that if $\mathfrak{p}$ is infinite then $\mathbb{Q}/\mathbb{Z}$ has to be replaced by $\frac{1}{2}\mathbb{Z}/\mathbb{Z}$ or 0 according to whether $\mathfrak{p}$ is real or complex. By the Local-Global Principle this global invariant map is *injective*. In other words: *every algebra A in $Br(K)$ is uniquely determined by its Hasse invariants (up to similarity).* In order to describe the structure of $Br(K)$ one has to describe the image of the invariant map. In other words: What are the conditions that a given system of rational numbers, $r_{\mathfrak{p}}$ for each prime $\mathfrak{p}$ of K, is the system of Hasse invariants of some algebra $A \in Br(K)$? The following conditions are evident:

1. There are only finitely many $\mathfrak{p}$ with $r_{\mathfrak{p}} \not\equiv 0 \bmod \mathbb{Z}$.

2. For an infinite $\mathfrak{p}$, we have $r_{\mathfrak{p}} \equiv \begin{cases} 0 \text{ or } \frac{1}{2} & \bmod \mathbb{Z} \quad \text{if } \mathfrak{p} \text{ is real.} \\ 0 & \bmod \mathbb{Z} \quad \text{if } \mathfrak{p} \text{ is complex.} \end{cases}$

Apart from this there is only one further condition, expressed in the following theorem.

Global Structure Theorem. (i) *For any central simple algebra A over the number field K, the sum formula for its Hasse invariants holds:*

$$\sum_{\mathfrak{p}} \left(\frac{A}{\mathfrak{p}} \right) \equiv 0 \bmod \mathbb{Z}.$$

(ii) *If $r_{\mathfrak{p}}$ is an arbitrary system of rational numbers (subject to the conditions 1. and 2. above) and if $\sum_{\mathfrak{p}} r_{\mathfrak{p}} \equiv 0 \bmod \mathbb{Z}$, then there is a unique $A \in Br(K)$ such that $\left(\frac{A}{\mathfrak{p}} \right) \equiv r_{\mathfrak{p}} \bmod \mathbb{Z}$ for each $\mathfrak{p}$.*

Today we would express this theorem by saying that the sequence of canonical maps

$$0 \to Br(K) \xrightarrow{\text{inv}} \oplus \sum_{\mathfrak{p}}{}' \mathbb{Q}/\mathbb{Z} \xrightarrow{\text{add}} \mathbb{Q}/\mathbb{Z} \to 0$$

is *exact*. This describes the Brauer group $Br(K)$, if considered via the map "inv" as a subgroup of the direct sum, as being the kernel of the map "add" which adds the components of the direct sum.

Although the Global Strucure Theorem uses the Local-Global Principle and is built on it, it is by no means an easy consequence of it. Perhaps this is the reason why the theorem is *not* treated in the Brauer-Hasse-Noether paper [BrHaNo:1932]; recall that this paper had to be written in haste. Nevertheless, as said above already, the authors of [BrHaNo:1932] stressed the point that their Local-Global Principle is of fundamental importance for the structure of the Brauer group.

The Global Structure Theorem, at least its first part (i), is in fact *equivalent to Artin's General Reciprocity Law* of class field theory. Hasse's proof can be found in [Has:1933], and it is the end point of a historic string of events stretching over several years since 1927. Let us briefly sketch chronologically the highlights in this development.

1927: Artin succeeded in [Art:1927] to prove his General Reciprocity Law which he had conjectured since 1923. Given an abelian extension $L|K$ of number fields, Artin's theorem established an isomorphism between the group of divisor classes attached to $L|K$ in the sense of class field theory, and the Galois group G of $L|K$. This isomorphism is obtained by associating to every prime $\mathfrak{p}$ of K which is unramified in L, its Frobenius automorphism $\varphi_{\mathfrak{p}} \in G$. This theorem has been said to be "the coronation of Takagi's class field theory".

Even before the appearance of his paper [Art:1927], Artin informed Hasse about his result and its proof. There followed an intense exchange of letters between Hasse and Artin discussing the consequences of Artin's Reciprocity Law. Already in his first such letter dated July 17, 1927, Artin mentioned that probably Hilbert's version of the reciprocity law may now be proved in full generality. Later he asked Hasse whether he could do it and Hasse agreed.

Accordingly, Hasse published in the same year 1927 a supplement to Artin's Reciprocity Law [Has:1927b] where (among other things) the product formula for the general m-th Hilbert symbol $\left(\frac{a,b}{\mathfrak{p}}\right)_m$ was established.

Here we do not intend to describe the definition of the Hilbert symbol; let it be sufficient to say that it serves to decide whether a given number $a \in K$ is an m-th power modulo a prime $\mathfrak{p}$, and that the product formula includes as a special case the Kummer m-th power reciprocity law together with its various supplementary laws.

The definition and management of the Hilbert symbol requires that the m-th roots of unity are contained in the base field K. Now Artin, in another letter to Hasse dated July 21, 1927, asked whether it would be possible to define some kind of Hilbert symbol *without* assuming that the proper roots of unity are contained in K. Hasse succeeded with this in 1929.

1929: Hasse's paper [Has:1930a] appeared in 1930 but since it had been received by the editors on March 7, 1929 already we count it for 1929. In this paper Hasse defined for an arbitrary abelian extension $L|K$ and $a \in K^\times$, for each prime $\mathfrak{p}$ of K the "norm symbol" $\left(\frac{a, L|K}{\mathfrak{p}}\right)$ as an element of the Galois group of $L|K$. More precisely, it is an element of the $\mathfrak{p}$-adic decomposition group $G_\mathfrak{p} \subset G$. This symbol assumes the value 1 if and only if a is a norm in the local extension $L_\mathfrak{p}|K_\mathfrak{p}$. The norm symbol is in some sense a generalization of the Hilbert symbol. If $\mathfrak{p}$ is unramified in L and $a = \pi$ a prime element for $\mathfrak{p}$ then $\left(\frac{\pi, L|K}{\mathfrak{p}}\right)$ equals the Frobenius automorphism $\varphi_\mathfrak{p}$ appearing in the Artin map. Hasse's norm symbol satisfies the product formula:

$$\prod_\mathfrak{p} \left(\frac{a, L|K}{\mathfrak{p}}\right) = 1$$

where the 1 on the right hand side denotes the neutral element of the Galois group. Hasse obtained this product formula from Artin's Reciprocity Law. Conversely, Artin's Reciprocity Law may be deduced from the above product formula.

As a side remark we mention that through this definition of the norm symbol Hasse discovered *local class field theory*. See [Has:1930b].

1931: On May 29, 1931 Hasse submitted his American paper [Has:1932a] to the Transactions of the AMS. In that paper he presented a comprehensive treatment of cyclic algebras over number fields. This was prior to the discovery of the Main Theorem, so Hasse did not yet know that every central simple algebra over a number field K is cyclic. For a cyclic algebra $A = (L|K, S, a)$ he compared the Hasse invariant $\left(\frac{A}{\mathfrak{p}}\right)$ with the norm symbol $\left(\frac{a, L|K}{\mathfrak{p}}\right)$. It turned out that the product formula of the (multiplicative) norm symbol provides the key for the proof of the sum formula for the (additive) Hasse invariant as stated in part (i) of the Global Structure Theorem.

Actually, the sum formula was not yet explicitly written down in Hasse's American paper; this was done in his next paper [Has:1933] only. But all the necessary ingredients and computations can be found in Hasse's American paper [Has:1932a] already. Although that paper had not yet appeared when the Brauer-Hasse-Noether paper [BrHaNo:1932] was written, the content of [Has:1932a] was known to Brauer and Noether too since Hasse had informed them about his results.

1932: In March 1932 Hasse sent his dedication paper [Has:1933] to Emmy Noether. This paper does not only contain a new proof arrangement for the Main Theorem, as we had reported earlier. In addition, Hasse stated and proved explicitly the Global Structure Theorem, not only Part (i) (which tacitly was already contained in [Has:1932a]) but also Part (ii). Moreover, the paper went well beyond Hasse's former papers in as much as now he *did not use Artin's Reciprocity Law in proving the Global Structure Theorem.*

We note that all three foregoing papers of Hasse, those of 1927, 1929 and 1931, were built on Artin's Reciprocity Law because the definition of the local norm symbol $\left(\frac{a,L|K}{\mathfrak{p}}\right)$ depended on Artin's global law. But now, in the Noether dedication paper, Hasse was able to use his invariants $\left(\frac{A}{\mathfrak{p}}\right)$ of algebras, for a purely local definition of the norm symbol $\left(\frac{a,L|K}{\mathfrak{p}}\right)$. This had been suggested to him by Emmy Noether who on a postcard of April 12, 1931 wrote the following. This letter was the reaction of Noether to a report of Hasse about his results in his American paper [Has:1932a].

> *I have read your theorems with great enthusiasm, like a thrilling novel; you have got really very far! Now ... I wish to have also the reverse: direct hyper-complex foundation of the invariants ... and thus hypercomplex foundation of the reciprocity law! But this may take still some time! Nevertheless you had done, if I remember correctly, the first step already in your skew field paper with the exponents e_p ?*

The "skew field paper" which Noether mentions, is Hasse's [Has:1931] which we had discussed above in Section 6.1. [45] As we have explained there, the Local Structure Theorem indeed can be used to provide a local definition of the Hasse invariant. Thus Noether had seen clearly the potential of this for her plan to reverse the argument, so that one first proves the sum formula of the Global Structure Theorem, and then interpret this as the product formula for Hasse's norm residue symbol $\left(\frac{a,L|K}{\mathfrak{p}}\right)$. The latter is equivalent to Artin's Reciprocity Law.

And Hasse followed Noether's hint and succeeded to give what Noether called a "hypercomplex proof of Artin's Reciprocity Law". Thus a close connection between the theory of algebras and class field theory became visible.

---

[45] See also Section 6.3.1 below.

While Artin's paper [Art:1927] with his reciprocity law had been named as the "Coronation of class field theory", similarly Hasse's paper [Has:1933] could now be regarded as the "Coronation of the theory of algebras".

We can now understand Artin's exclamation which we have cited in Section 2, namely that he regards this as the *greatest advance in Number Theory of the last years...* " When Artin wrote that letter in November 1931, Hasse's paper [Has:1933] was not yet out. However Artin seemed to have clearly seen, as Emmy Noether had done, the potential of the Local-Global Principle as a foundation of class field theory. In fact, in his letter he continued:

> *At present I am giving a course on class field theory, and next semester I will continue by becoming hypercomplex...*

Thus Artin intended to discuss the theory of hypercomplex systems, i.e., algebras, with the view of its application to class field theory. [46]

6.3 Remarks

6.3.1 Arithmetic of Algebras and Hensel's Methods

Twice in our discussion we had occasion to refer to Hasse's paper [Has:1931] on local algebras. The first time this was in Section 5.1 when we reported that the local splitting theorem was an almost immediate consequence of the results of that paper. The second time was in Section 6.1 when we discussed the local structure theorems. In both situations we have seen that Hasse's paper [Has:1931] contained the fundamental ingredients which led to success.

We note that this paper [Has:1931] was received by the editors on June 18, 1930 already, long before the Brauer-Hasse-Noether paper was composed, and even before Hasse had formulated the conjecture of the Main Theorem in a letter to Noether (see Section 7.2, p. 56). In fact, the original motivation for Hasse to write this paper was *not* directly connected with the Main Theorem. From the introduction of [Has:1931] we infer that Hasse regarded his paper as a new approach to understand the *arithmetic of algebras*, based on the ideas of Hensel, in the same manner as he had applied those ideas to the investigation of the arithmetic of commutative number fields.

---

[46] The "next semester" was the summer term 1932. In that semester Artin gave a course with the title "Algebra", and he presented there the algebraic theory of hypercomplex systems, i.e., algebras. Lecture Notes for this course had been taken down by the student Ernst August Eichelbrenner, and a copy is preserved. However from these notes it appears that Artin covered the algebraic theory of algebras over an arbitrary field only, but not the special situation when the base field is a number field. In particular, the connection to class field theory is not mentioned. But it may well have been that Artin covered those more advanced topics in a special seminar parallel to this course.

In this paper we cannot give a comprehensive account of the development of the arithmetic of algebras during the 1920s; this is an exciting story but would need much more space than is available here. The following brief comments should help to put Hasse's paper [Has:1931] into the right perspective.

The study of the *arithmetic theory* of algebras had been started systematically by Dickson whose book [Dic:1923], entitled

Algebras and their Arithmetics,

had received much attention, in particular among German mathematicians. This book contained not only a complete treatment of the Wedderburn structure theorems for algebras, but also a systematic attempt to develop an *arithmetic theory of orders* of an algebra.

If A is any algebra over a number field K and $\mathbb{Z}_K$ the ring of its integers, an *order* R is defined to be a subring (containing 1) of A which is a finite $\mathbb{Z}_K$-module and generates A as a vector space over K. The "arithmetic" of R manifests itself in the structure of ideals of R. The arithmetic of such order becomes particularly lucid if the order is *maximal*, i.e., not properly contained in a larger order of the algebra. Perhaps it is not an exaggeration to say that the most important feature of Dickson's book was to give the definition of *maximal* orders of an algebra and to point out that the arithmetic of those maximal orders is particularly lucid – in the same way as in the commutative case, i.e., algebraic number fields K, where the maximal order $\mathbb{Z}_K$ and its prime ideal structure is the first object to study in algebraic number theory, whereas arbitrary orders, i.e., those which are not integrally closed, carry a more complicated ideal theory.

Now we observe that in Dickson's book, after maximal orders have been defined and their elementary properties developed, they are in fact not treated in full generality. The discussion is largely restricted to very special cases, namely when there exists a euclidean algorithm. One knows in algebraic number theory that such cases are rare.

The first who set out to remedy this unsatisfying situation was Andreas Speiser in his paper [Spe:1926]. Also, he arranged for a German translation of Dickson's book [Dic:1927] and included his paper as an additional chapter. [47]

But still, Speiser's treatment was only the beginning. Soon Artin published a series of three seminal papers in which the arithmetic of maximal orders was developed in full [Art:1928a], [Art:1928b], [Art:1928c]. The second of these papers contained the generalization of Wedderburn's structure theorems to what

---

[47] The translation had been done by J. J. Burckhardt who recently had his 100th birthday in good health. See [Fre:2003]. – Actually, the German edition is not merely a translation of the American book. Dickson had presented a completely reworked manuscript for translation. The book had been reviewed by Hasse [Has:1928] in the *Jahresbericht der DMV*.

today are called "Artinian rings", i.e., rings with minimum condition for ideals. [48] This is necessary if one wishes to study the structure of the residue class rings of a maximal order with respect to arbitrary two-sided ideals which are not necessarily prime. The third of Artin's papers then developed the ideal theory of maximal orders of a simple algebra, in complete analogy to Noether's theory of Dedekind rings (which had just been published the year before). The non-commutativity implies that the ideals do not necessarily form a group but (with proper definition of multiplication) a so-called "groupoid" in the sense of Brandt [Bran:1930].

It was Speiser's work and, in addition, this series of papers by Artin which had inspired Hasse to write his paper [Has:1931] on local division algebras. In his introduction he refers to Speiser and Artin, and says:

> *I am extending the original idea of Speiser in the same sense as Hensel's arithmetic of number fields had done with the original idea of Kummer. For, if one considers residue classes with respect to arbitrary powers of a prime ideal* $\mathfrak{p}$ *then this means to extend the field of coefficients* $\mathfrak{p}$*-adically. In this way, Speiser's residue classes modulo* $\mathfrak{p}^s$ *(for arbitrary s) are replaced by an algebra over the* $\mathfrak{p}$*-adic completion...*
>
> *In this way it is possible to build the hypercomplex arithmetic in a surprisingly simple way.*

And as part of this program, Hasse mentions:

> *Moreover I have succeeded in giving a complete description of all existing division algebras with a complete* $\mathfrak{p}$*-adic field as its center, and of its arithmetic and algebraic structure – in analogy to the well-known fact that with the real number field as center there exists only one skew field, the ordinary quaternion field.*

Thus Hasse's local structure theorems which we had cited from [Has:1931a], constituted only one aspect of this paper. The other and broader one was to build non-commutative arithmetic in maximal orders of algebras, by using Hensel's ideas of localization.

In this light we can understand why Hasse in his dedication text of the Brauer-Hasse-Noether paper had mentioned "Hensel's *p*-adic methods" as being responsible for the success.

---

[48] At that time it was not yet known that the minimum condition for ideals implies the maximum condition. Hence Artin in his paper required the validity of the minimum as well as the maximum condition.

6.3.2 Class Field Theory and Cohomology

In the section "applications" of the Brauer-Hasse-Noether paper there is a subsection concerning

Generalization of central theorems of class field theory to the case of arbitrary Galois extensions of number fields.

The problem is the following: Ordinary class field theory in the sense of Takagi refers to *abelian* extensions of number fields. The abelian extensions L of a number field K are characterized by certain groups, called "ray class groups" of divisors, which are constructed within the base field K – in such a way that the decomposition type in L of primes $\mathfrak{p}$ of K can be read off from the behavior of the primes in the corresponding ray class group. Question: Is a characterization of a similar kind possible for Galois extensions of K which are not necessarily abelian? It was known since Hasse's class field report [Has:1926] that this is not possible by means of ray class groups; this is the content of what Hasse [Has:1933a] called *"Abgrenzungssatz"* (theorem of delimitation). But there may be other groups or objects which are defined within K and can serve to describe Galois extensions $L|K$.

Now, in the Brauer-Hasse-Noether paper, Hasse proposes to use Brauer groups. He shows that every Galois extension $L|K$ of number fields is uniquely determined by its Brauer group $Br(L|K)$. And the decomposition type in L of a prime $\mathfrak{p}$ of K can be read off from the $\mathfrak{p}$-adic behavior of the elements in $Br(L|K)$. The proof is almost immediate using the splitting theorems of Section 5.1 together with well known density theorems of algebraic number theory.

In consequence there arises the problem how to describe the Brauer groups $Br(L|K)$ within K without resorting to L. This problem is not treated in the Brauer-Hasse-Noether paper. However it has stimulated several mathematicians, including Artin and Noether, to look more closely into the Brauer group or, equivalently, into the group $H^2(L|K)$ of factor systems.

In the Artin-Hasse correspondence we find 5 letters between March and May 1932 where Artin tries to give congruence criteria for the decomposition type of a prime $\mathfrak{p}$ of K in L by means of factor systems. However his results were disappointing to him. He wrote to Hasse:

In the non-abelian case we just obtain the old method to apply class field theory to subfields over which the whole field is cyclic... This is only a somewhat beautified combination of the ordinary class field theory ... I have the impression that something completely new has to be added.

Hasse seems to have been more optimistic. In his talk [Has:1932c] at the International Congress of Mathematicians, Zürich 1932, he said, after having reported about the Main Theorem:

Further work of Artin, E. Noether and myself has shown that Theorem 3 (and that method in general) is a powerful tool to deal with the great question in the center of modern number theory, the decomposition law in general Galois number fields.

Here, "Theorem 3" means the Main Theorem, and "that method" had been explained in Hasse's text before, namely:

... a combination of the arithmetic methods of Hensel, which I have carried into this theory following Speiser, with certain algebraic methods which, based on earlier investigations of Speiser and I. Schur, have recently been developed by R. Brauer and E. Noether. [49]

Emmy Noether seems to have steered a middle line. On the one hand she was informed about Artin's unsuccessful attempts. In her invited address [Noe:1932] at the Zürich Congress she says the following, after having reported on Hasse's proof of the reciprocity law by means of algebras, and on some further developments of Chevalley [Che:1933] about factor systems:

... At the same time I have to qualify this by saying that the method of crossed products does not seem to cover the full theory of Galois extensions of number fields. This is a consequence of new, still unpublished results of Artin which are based on Hasse's methods...

On the other hand, Emmy Noether in her own work pushes the computation with factor systems further along by what she calls the "Principal Genus Theorem" (*Hauptgeschlechtssatz*). Seen from today's viewpoint, her work is of cohomological nature and her Principal Genus Theorem is essentially the vanishing of the 1-cohomology of the idèle class group, or at least of some finite level of it.

In the course of later developments the idea of approaching class field theory for Galois extensions by means of factor systems has been dropped. The new concept for an edifice of class field theory for Galois extensions, due to Langlands, looks quite different.

But the extensive computations with factor systems have had a significant consequence in the long term, namely the rise of algebraic cohomology and its application in ordinary class field theory (i.e., for abelian extensions). While Hasse had introduced simple algebras into class field theory in [Has:1933], these have survived in modern times as 2-cohomology classes only. Accordingly the

---

[49] It may seem strange that Hasse did not mention Albert who also had an independent share in the proof of the Main Theorem (see Section 8.) We can only speculate about the reason for this (if there was any particular reason at all). It may have been the fact that, for one thing, Hasse had mentioned Albert earlier in his text together with Dickson and Wedderburn, and that on the other hand the methods which he refers to in the present connection are concerned with computations on factor systems in the realm of class field theory, which indeed cannot be found in Albert's papers.

exact sequence which we have written down in Section 6.2 (p. 42) immediately after the Global Structure Theorem, is now written in the form

$$0 \to H^2(K) \longrightarrow H^2(I_K) \longrightarrow \mathbb{Q}/\mathbb{Z} \to 0 \,,$$

I_K being the idèle group (suitably topologized) and $H^2(I_K) = \oplus \sum_{\mathfrak{p}} H^2(K_{\mathfrak{p}}^{\times})$.

Many years later, Artin and Tate presented in their 1952 Seminar Notes [ArTa:1968] an axiomatic foundation of class field theory. Their axioms were given in the language of cohomology which by then was already well developed. There are two main axioms. Their Axiom I is essentially the cohomological version of the exactness of the above sequence at the term $H^2(K)$. And their Axiom II is essentially equivalent to the exactness at $H^2(I_K)$.

We have mentioned all this in order to point out that it started with the Brauer-Hasse-Noether paper [BrHaNo:1932].

The Team: Noether, Brauer and Hasse

As seen above, there was a close collaboration between Brauer, Hasse and Noether which finally led to the Main Theorem. We have tried to find out how this cooperation started and developed.

7.1 Noether's Error

We do not know when Emmy Noether and Richard Brauer had met for the first time. There is a letter from Noether to Brauer dated March 28, 1927 which seems to be a reply to a previous letter from Brauer to her. [50] The letter is published and discussed in the beautiful book of C. Curtis *"Pioneers of representation theory"* [Cur:1999], p.226; it starts as follows:

> *Dear Mr. Brauer! I am very glad that now you have also recognized the connection between representation theory and the theory of noncommutative rings, the "algebras", and the connection between the Schur index and division algebras.*

The tone of the letter is somewhat like that from an instructor to a young student [51], giving him good marks for success in his studies. But then she becomes serious:

> *In regard to these fundamentals our investigations are, of course, in agreement; but then it seems to me there is a divergence.*

And she continues to describe this divergence, followed by an essay on how she likes to view the situation, with the unspoken invitation that Brauer too should take the same viewpoint.

The subject is representation theory, the Schur index and splitting fields. Noether advocates that the whole theory be subsumed under the theory of algebras. On this project she had been working already for some time. In the winter

---

[50] In the same year 1927 Brauer did his *"Habilitation"* at the University of Königsberg. It seems that Brauer had sent his thesis (*Habilitationsschrift*) to Noether asking for her comments, and that the above mentioned letter is Noether's reply. Brauer submitted his paper [Bra:1928] to the *Mathematische Zeitschrift* on July 22, 1927, four months after Noether's letter.

[51] Noether was 19 years older than Brauer, who was 26.

semester 1924/25 she had given a course on this subject. [52] In September 1925 she had given a talk at the annual meeting of the DMV at Danzig with the title "Group characters and ideal theory". [53] In the abstract of this talk [Noe:1925] she writes:

> *Frobenius' theory of group characters – i.e., representation of finite groups – is seen as the ideal theory of a completely reducible ring, the group ring.*

She continues with the Wedderburn structure theorems for algebras and how these are to be interpreted within representation theory. She ends up with the sentences:

> *Thus the theory of Frobenius is subsumed. A more detailed presentation is to be published in the* Mathematische Annalen.

We see that already in 1925 Noether had a clear view of what was necessary to develop representation theory within the framework of algebras. But the promised publication had to wait for quite a while. Noether was not a quick writer; more often her ideas went into the papers of other people rather than forcing herself to write a manuscript. It is not clear which paper she was announcing here; some years later there are two publications of Noether on representation theory [Noe:1929] and [Noe:1933] (both in the *Mathematische Zeitschrift* and not in the *Mathematische Annalen*). The first of these papers [Noe:1929] consists essentially of the notes taken by van der Waerden of Noether's lecture in the winter semester 1927/28. These lecture notes have been said to constitute "one of the pillars of modern linear algebra". [54] The second paper [Noe:1933] is somewhat more closely related to the topics of her correspondence with Brauer. We have the impression that in the announcement she had in mind one longer paper but in her hands this became too long and thus was divided into two parts.

Returning to Noether's letter to Brauer on March 28, 1927: In one of the statements in that letter she claims that each minimal splitting field of a division algebra D is isomorphic to a maximal commutative subfield of D. But this turned out not to be true. There do exist minimal splitting fields of D whose degree is larger than the index; an embedding into D is possible if and only if *the degree* of the splitting field is minimal, i.e., equals the index of D.

We will see that this error had important consequences, leading to the Brauer-Hasse-Noether theorem.

---

[52] I am indebted to Mrs. Mechthild Koreuber for showing me her list of the Noether lectures in Göttingen 1916-1933, copied from the *"Vorlesungsverzeichnis"*. For the winter semester 1924/25 we find the announcement of a 4 hour lecture on group theory. This seems to be the lecture where Noether first expounded her ideas of doing representation theory within the framework of algebras.

[53] This was the same meeting where Hasse gave his famous report on class field theory. Noether's and Hasse's talks were scheduled at the same session. See [Roq:2001].

[54] Cited from [Cur:1999] who in turn refers to Bourbaki.

We do not know whether Brauer had replied to Noether's letter. But we do know that both met at the next annual meeting of the DMV, on September 18–24, 1927 at the spa of Kissingen. Neither Noether nor Brauer were scheduled for a report at that meeting but certainly they talked about the topic of Noether's letter. From the correspondence over the following weeks we can obtain a fairly good picture of what they had discussed in Kissingen. Apparently Brauer knew that Noether's claim cited above was erroneous, i.e., that there could exist minimal splitting fields whose degree is larger than the index. And he told Noether. But then Noether asked whether the degrees of the minimal splitting fields may be bounded. This too seemed to be doubtful. Noether wished to check the question at the smallest example, namely the ordinary quaternions over the rational number field $\mathbb{Q}$.

In Kissingen they could not settle the question. Two weeks later, on October 5, 1927, Noether sent a postcard to Hasse (with whom she had corresponded since 1925):

> *Dear Mr. Hasse! Can you tell me whether the general existence theorems for abelian fields yield the following: For every n there exists al least one (perhaps arbitrary many) cyclic field over the rational number field of degree 2^n such that its subfield of degree 2^{n-1} is real, and it admits a representation of (-1) by at most three squares (squares of fractional numbers)...* [55]

If such fields would exist then there would exist minimal splitting fields of arbitrary large degree. Noether continued:

> *R. Brauer conjectured in Kissingen the non-boundedness* [of the degrees of minimal splitting fields]*, but his examples were more complicated than quaternion fields. It would follow that one knows much less about those minimal fields than I had believed for some time.*

Hasse reacted immediately. Already on the next day, on Oct 6, 1927, he sent to Noether a 4-page manuscript in which he gave a detailed proof that indeed, such fields do exist. For us it is of interest that his proof was based upon the *Local-Global Principle*: not the Local-Global Principle for algebras (this was not known yet) but for quadratic forms, which Hasse had discovered 1922 in his famous thesis [Has:1923]. The quadratic form relevant to Noether's question is the sum of four squares: $f(x) = x_1^2 + x_2^2 + x_3^2 + x_3^2$. The question whether $f(x)$ has a non-trivial zero in a field is, of course, identical with the question whether the field is a splitting field of the ordinary quaternions.

---

[55] Of course, the condition that the subfield of degree 2^{n-1} should be real, is always satisfied and hence could be omitted. This is another instance where we can see that Noether often wrote her postcards very impulsively and dispatched them immediately – without thinking twice about the text. (Very much like some people send e-mail messages nowadays.) If she had, she would certainly have noticed that the condition to be real is superfluous, as she admits in her next postcard.

So we see here, in Hasse's letter of October 1927, the nucleus of what in 1931 would become the Local-Global Principle for algebras in the Brauer-Hasse-Noether paper.

Immediately thereafter, on Oct 10, 1927, Noether wrote to Brauer sending him Hasse's letter which solves the question under discussion, and proposing a joint note, to be published together with a note of Hasse. There followed a series of letters within the triangle Brauer-Hasse-Noether, discussing details about the planned notes and possible generalizations. After a while Brauer succeeded to construct those fields without using Hasse's Local-Global Principle for quadratic forms. Hasse asked Brauer to explain to him the group theoretic relevance of his example, which Brauer did in full detail.

Finally there appeared a joint note of Brauer and Noether [BrNo:1927], and immediately after it in the same journal a note by Hasse [Has:1927a]. The offprints of both papers were bound together and distributed in this form.

Here we see the first instance where Brauer, Hasse and Noether had formed a team towards a common goal – as a consequence of Noether's error concerning minimal splitting fields.

7.2 Hasse's Castles in the Air

Thereafter Hasse became increasingly interested in the theory of algebras because he had seen that number theory, in particular class field theory and the local p-adic theory, could be used there profitably. Brauer became interested in class field theory because of the same reason. And Noether, who had brought the two together in the first place, was pleased because she observed that "her methods" were accepted by both.

There resulted a regular exchange of letters and information between the three members of the team. Brauer learned from Hasse about class field theory and Hasse learned from Brauer about algebras and group rings.

For instance, upon a request of Hasse, Brauer wrote him on July 9, 1929 all that he knew about group rings in the setting of algebras. On October 26, 1929 Brauer sent to Hasse his notes which he had composed for his lecture course at Königsberg. On March 16, 1930 Hasse wrote to Brauer explaining in detail his main ideas and results on skew fields over $\wp$-adic fields (they appeared later in [Has:1931]). We have discussed that paper in Section 6.1 in connection with the Local Structure Theorem. (See also Section 6.3.1.) At the end of his letter Hasse observed that over a local field $K_\mathfrak{p}$ (with $\mathfrak{p}$ finite) there exist extension fields with non-cyclic Galois group; nevertheless every central skew field over $K_\mathfrak{p}$ is cyclic. And then he asks Brauer:

> *Now I would like to ask whether you know a central skew field over the rational or an algebraic number field which does not admit an abelian, or at least no cyclic, splitting field which is a maximal commutative subfield? Is it*

true that the direct product of two quaternion algebras, which you mentioned the other day, is non-cyclic in this sense?

We see that Hasse contemplates about whether globally every central skew field is abelian or perhaps even cyclic, i.e., the Main Theorem. But he is not sure and wants to know the opinion of Brauer.

In his reply on April 18, 1930 Brauer thanked Hasse for this letter, saying that he was highly interested in Hasse's beautiful results on the arithmetic of hypercomplex numbers. Concerning Hasse's question he wrote:

Unfortunately I am not able yet to answer your question. I do not even know whether there exist skew fields (of finite rank over their center) which do not admit a normal [56]field as a maximal subfield. Formerly I have tried without success to prove the existence of such a normal field. Now I am trying the opposite, to construct an example which does not admit such a normal field.

But, he adds, although he has an idea how to construct such an example, this will probably be very sophisticated. Apparently he did not succeed. To our knowledge, such an example was first given by Amitsur [Ami:1972]. But, of course, the base field in Amitsur's construction was not an algebraic number field. [57] Brauer continues:

Also your more specific question whether there are skew fields which are not of Dickson type (i.e., which have no cyclic maximal subfields), I am not able to answer. The product of two quaternion algebras which you mentioned, is not eligible since its center is not algebraic...As soon as I will know more about it I will write again to you.

The said "product of two quaternion algebras" had been treated by Brauer in his paper [Bra:1930] over the rational function field $\mathbb{Q}(u, v)$ of two variables. The examples provided by Brauer have exponent 2 and index 4 but the question whether they are cyclic is not treated in [Bra:1930]. The first who explicitly constructed a non-cyclic division algebra was Albert [Alb:1932a]. [58]

At the end of his letter Hasse said that he would very much like to talk personally to Brauer about these questions. And he announced that in the fall he will be in Königsberg (the place where Brauer lived) for the meeting of the

---

[56] A "normal" field extension, in the terminology of the time, means a Galois extension.

[57] Since then there has developed an extensive literature trying to understand the construction of non-crossed product division algebras. As a noteworthy example we mention the work by Brussel who uses Wang's counterexamples to Grunwald's theorem to construct non-crossed products. See [Bru:1997], and also subsequent papers of the same author.

[58] Albert used a construction similar to but not identical with Brauer's. Note that in a footnote on the first page of [Alb:1932a] it is claimed that Brauer's construction was false, but at the end the author admitted in an additional note that the difficulty was one of the interpretation of language, rather than a mathematical error.

DMV (German Mathematical Society). We may safely assume that Hasse and Brauer had a very thorough discussion there, together with Emmy Noether who also participated at the meeting. Unfortunately there is no record about their conversations. [59]

In December 1930 Hasse seemed to have made up his mind and written up some coherent conjectures about the structure of algebras over number fields. He did so in a letter to Emmy Noether. We do not know Hasse's letter but we do know the reaction of Emmy Noether. From that we can conclude that among Hasse's conjectures was the Main Theorem and the Local-Global Principle, but also the consequence that over number fields, the index of a central simple algebra equals its exponent. It appears that Hasse had mentioned that his conjectures do not yet have a solid foundation. For Noether replied on December 19, 1930 :

> *Yes, it is a terrible pity that all your beautiful conjectures are floating in the air and are not solidly fixed on the ground: for part of them – how many I do not yet see – hopelessly crash through counterexamples in a very new American paper... by Albert. From that it follows, firstly, that the exponent can indeed be smaller than the index, already with the rational number field as center, and furthermore that your theory of forms cannot be transferred to forms of higher degree. Whether your conjecture concerning cyclic splitting fields holds is at least doubtful.*

After this, Noether proceeds to explain to Hasse the counterexamples which she purports to have found in Albert's paper [Alb:1930]. And she ends the letter by asking Hasse to inform her if his cyclic splitting field is crashed too; this is an indication that indeed the Main Theorem was conjectured by Hasse, as we had stated above already.

We see that Noether did not yet believe in the validity of the Main Theorem of which one year later she would be a co-author.

In her letter Noether had mentioned Hasse's "theory of forms" (*Formentheorie*). We do not know precisely what was meant by this. In the present context it seems probable that "forms" were to be understood as norm forms from central simple algebras, and that Hasse had the idea that some kind of Local-Global Principle should hold for those norm forms – in analogy to quadratic forms which he had treated in his thesis and following papers. If our interpretation is correct then we can conclude that here, in December 1930, Hasse's conjectures included the Local-Global Principle for algebras.

---

[59] Except Noether's reference to their trip to Nidden of which she reminded Hasse in her letter of Nov 12, 1931 which we have cited in Section 4.3. – As a side remark we would like to draw the reader's attention to the essay [Tau:1979] by Olga Taussky-Todd who reported about her experiences during this Königsberg meeting, including her vivid recollection of how Noether and Hasse seemed to have a good time discussing her (Olga's) results on the capitulation problem of class field theory.

The vivid language Noether had used in her reply appears to be quite typical of her style. Hasse apparently did not mind it. He seems to have checked Albert's paper cited by Noether and found out that Noether's interpretation of Albert's result was incorrect. And he wrote this to her. Whereupon on December 24, 1930 (Christmas eve!) she replied:

> *This letter is a* pater peccavi. *For, your castles in the air are not yet crashed...Your counterexample has cleared up the situation for me.*

Noether said that she had extracted from Albert's paper the opposite of what was in it. After longish explanations of her error she writes:

> *After all, it appears probable that with an algebraic number field as center, exponent and index do always coincide.*

From then on, Noether was on Hasse's side and she vividly advocated his conjectures. She arranged that Hasse was invited to give a colloquium talk in the Göttingen Mathematical Society on January 13, 1931. The title of his talk was "On skew fields". Hasse's manuscript for this talk is preserved. This shows:

> *In Göttingen, on January 13, 1931 Hasse publicly announced his conjecture for the Main Theorem.* [60]

Although Hasse in his manuscript did not say anything about how he would try to approach this conjecture, implicitly we can see the Local-Global Principle behind it. For, he reports extensively on his results about local algebras [Has:1931], and that he had discovered they are always cyclic. And starting from the local results he formulates the conjectures for the global case.

7.3 The Marburg Skew Congress

Some weeks later Hasse and Noether met again, this time in Marburg on the occasion of a small congress, today we would say *workshop*, on skew fields. [61] Hasse's idea was to bring together mathematicians who were active in the theory of algebras, of class fields and of group representations, in order to join forces with the aim of solving his conjectures. It seems that he had discussed this plan with Emmy Noether when he visited Göttingen for his colloquium talk in January; they had agreed to have the workshop at the end of the winter semester,

---

[60] In the book [Cur:1999] it is said that the Main Theorem had been conjectured by Dickson already. But we did not find this in Dickson's works, and after inquiring with the author he replied: *"The statement in my book about Dickson's conjecture has to be withdrawn."* His statement was based on an assertion in Feit's obituary article on Brauer [Fei:1979] which he had accepted without checking.

[61] Starting with the summer semester 1930 Hasse had accepted a professorship in Marburg as the successor of Kurt Hensel. On this occasion he had been granted some funds for inviting visitors to Marburg for colloquium lectures etc.

February 26 to March 1, 1931. Noether liked to call the meeting the "skew congress".

Of course Richard Brauer was invited too. In the letter of invitation to Brauer, dated February 3, 1931, Hasse said this will become a small, *"gemütlicher"* congress on skew fields. [62] Hasse mentions the names of the other people who were invited: Noether, Deuring, Köthe, Brandt and Archibald (disciple of Dickson). Moreover, Hasse wrote, invitations will be sent to all people who are interested, e.g., Artin, Speiser, I. Schur – but the funds available to Hasse were not sufficient to cover the expenses of all of them. From the correspondence of Hasse with Krull we know that Krull was invited too, as well as F. K. Schmidt (both in Erlangen) but they were unable to come because at precisely the same day they had invited a guest speaker to the Erlangen colloquium.

Emmy Noether, in her letter to Hasse dated February 8, 1931 offered a title for her own talk, and she forwarded already some proposals for the program of the skew congress. She proposed the talks of R. Brauer, Noether, Deuring, Hasse to be held in this order, so that every one of the lecturers could build on the foregoing talks. The other lectures, she wrote, were independent. In her letter she also mentioned Fitting, a Ph. D. student of hers. [63]

Moreover, Noether strongly recommended to invite Jacques Herbrand, her Rockefeller fellow. She wrote that Herbrand had worked on Logic and Number Theory only. Number Theory he had learned from Hasse's Class Field Report [Has:1927], [Has:1930] and Hasse's papers on norm residues. She recommends

---

[62] The following passage from Hasse's letter to Brauer may also be of interest: *"As I told you already last summer, we are doing representation theory this year, and we use mainly your lecture notes which you had kindly sent me."*

[63] Recently some letters from Emmy Noether to Paul Alexandroff have been published by Renate Tobies [Tob:2003], and there, in a letter of October 13, 1929, we read: *"Hasse will go to Marburg as the successor of Hensel; I wrote to him concerning connected visiting lectures but have not yet obtained a reply."* In her comments Tobies interpreted this as Noether having written to Hasse proposing to establish in Marburg something like the "skew congress". But this interpretation remains doubtful. Among the letters from Noether to Hasse of that time we did not find any passage of that kind. Instead, a few days earlier than her letter to Alexandroff she wrote a letter to Hasse (October 7, 1929) where we read something else. There she first congratulates Hasse for having received the offer from Marburg University, and then she proposed Alexandroff to be named as Hasse's successor in Halle. It is not clear how this blends with what she wrote to Alexandroff about "connected visiting lectures" (*zusammenhängende Gastvorlesungen*). We could speculate that she hoped, if Alexandroff would have been mentioned in the list of possible successors of Hasse in Halle, then at least she could obtain funds for inviting him to Germany for longer periods. But this is pure speculation and so the meaning of Noether's writing to Alexandroff remains in the dark, as for our present knowledge. – In any case, as we see, once Noether knew about Hasse's idea for this skew congress, she actively stepped in and helped him in the planning.

him as a mere participant only but if he is to give a talk too then he could report on his results about the integral representation of the Galois group in the group of units [Her:1930].

In those years it was not uncommon that colloquium lectures, meetings etc. of the foregoing year were reported in the *Jahresbericht der DMV*. Accordingly, in its 1932 issue we can find the Marburg 1931 skew congress program as follows: [64]

26. 2.	H. Hasse	Dickson skew fields of prime degree.
27. 2.	R. Brauer (Königsberg)	Galois theory of skew fields.
	M. Deuring (Göttingen)	Application of non-commutative algebra to norms and norm residues.
	E. Noether (Göttingen)	Hypercomplex structure theorems and number theoretic applications.
	R. Archibald (Chicago)	The associativity conditions in Dickson's division algebras.
	H. Fitting (Göttingen)	Hypercomplex numbers as automorphism rings of abelian groups.
28. 2.	H. Brandt (Halle)	Ideal classes in the hypercomplex realm.
	G. Köthe (Münster)	Skew fields of infinite degree over the center.

There may have been other participants in the workshop who did not deliver a talk. Perhaps Herbrand was one of them.

We see that Noether's proposals for the order of the talks were not realized; perhaps the mutual dependance of the talks was not so strong that a unique order would follow. The first day (Feb 26) was the day of arrival; we know that the visitors arrived late at noon and hence Hasse's talk had probably been scheduled some time in the afternoon. We observe that in the title he used the old terminology "Dickson algebras". Hasse's notes for this lecture are preserved and there, however, he speaks of "cyclic algebras". From the notes we infer that Hasse presented a report on cyclic algebras of prime degree, based on his earlier work [Has:1931] and the Hilbert-Furtwängler Norm Theorem (see Section 4.1). At the end he presented a number of problems, thus setting the pace for this meeting and also for future work in the direction of the Local-Global Principle (see Section 4) and the structure of the Brauer group (Section 6).

If Hasse's aim in this workshop had been to get a proof of the Main Theorem then this aim was not achieved. But there resulted a general feeling that the final solution was near. Hasse, in particular, seems to have been encouraged and kept himself busy working on the problem.

---

[64] We have found the reference to this program in Tobies' article [Tob:2003].

Already a week later, on March 6, 1931 Hasse proudly sent a circular to the participants of the skew congress with the following telegram style message:

Dear Sir/Madam: Just now I have proved the norm theorem in question for relatively cyclic fields, and more is not needed for the theory of cyclic division algebras.

In other words: He had proved the Local-Global Principle to hold for cyclic algebras of any index, not necessarily prime. This was one of the problems which he had stated in his talk on February 26. Hasse published the result in [Has:1931a], and its consequences for cyclic algebras were announced in [Has:1931b]. A complete theory of cyclic algebras over number fields followed in his American paper [Has:1932a].

From here on, we have already reported the further development in the foregoing Sections 4–6.

The American Connection: Albert

8.1 The Footnote

There is a footnote in the Brauer-Hasse-Noether paper [BrHaNo:1932] which reads as follows (in English translation):

> *The idea of reduction to solvable splitting fields with the help of Sylow's group theoretical theorem has been applied earlier already by R. Brauer, namely to show that every prime divisor of the index also appears in the exponent [Bra:1928]. Recently A. A. Albert has developed simple proofs for this idea, not dependent on representation theory, also for a number of general theorems of the theory of R. Brauer and E. Noether ([Alb:1931a], [Alb:1931c]; for the reduction in question see in particular Theorem 23 in [Alb:1931c]).*

> Added in proof. *Moreover A. A. Albert, after having received the news from H. Hasse that the Main Theorem has been proved by him for abelian algebras (see in the text below), has deduced from this, independent from us, the following facts:*
>
> *a) the Main Theorem for degrees of the form 2^e,*
> *b) the theorem 1 below (exponent = index),*
> *c) the basic idea of reduction 2, and also of the following reduction 3, naturally without referring to reduction 1, and accordingly with the result: for division algebras D of prime power degree p^e over Ω there exists an extension field Ω' of degree prime to p over Ω, so that $D_{\Omega'}$ is cyclic.*
>
> *Of course, all three results are now superseded by our proof of the main theorem which we have obtained in the meantime. But they show that A. A. Albert has had an independent share in the proof of the main theorem.*

> *Finally, A. A. Albert has remarked (after knowing our proof of the main theorem) that our central theorem I follows in a few lines from the theorems 13, 10, 9 of a paper which is currently printed ([Alb:1931d]). The proof of those theorems is based essentially on the same arguments as our reductions 2 and 3.*

Here, the "reductions" are the steps in the proof of the Main Theorem in the Brauer-Hasse-Noether paper. As explained in Section 4.2, "reduction 2" is Brauer's Sylow argument, "reduction 3" is Noether's induction argument in the solvable case.

This footnote has aroused our curiosity. We wanted to know more about the relation of Albert to Hasse, and about Albert's role in the proof of the Main Theorem. The correspondence between Albert and Hasse during those years is preserved. Our following report is largely based on these documents.

To have a name, we shall refer to this footnote as the "Albert-footnote". For later reference we have divided the Albert-footnote by horizontal lines into three parts. It will turn out that these parts were added one at a time. The horizontal lines are not contained in the original. Also, in the interest of the reader we have changed the references to Albert's papers in the original footnote to the corresponding reference codes for this article.

8.2 The First Contacts

A. Adrian Albert (1905–1972) had been a disciple of Dickson. We have said in Section 6.3.1 (p. 46) already that Dickson's book "Algebras and their Arithmetics" had a great influence on the work of German mathematicians in the 1920s. The results of Dickson and his disciples were noted carefully and with interest by the German mathematicians around Noether.

Already in Section 7 (p. 56) we have met the name of Albert. There we reported that Noether, after reading a recent paper by Albert, thought erroneously for a short time that, with the help of Albert's results, she could construct a counterexample to some of Hasse's conjectures in connection with the Main Theorem.

We find the name of Albert again in the manuscript which Hasse had written for his personal use on the occasion of the colloquium talk at the Göttingen Mathematical Society, January 13, 1931. This was the talk where Hasse publicly announced for the first time his conjecture for the Main Theorem; we have discussed it on p. 57. After having announced his conjecture, Hasse, according to his manuscript, reported what was known for division algebras D of small index n. For $n = 4$ he cited Albert for the fact that every division algebra contains a maximal subfield of degree 4 which is abelian with Galois group non-cyclic of type $(2, 2)$. But Hasse noted in his manuscript that Albert's proof cannot be valid because "*there exist, as is easily seen, cyclic division algebras of index* 4 *which do not contain an abelian subfield with group of type* $(2, 2)$."

We do not know what Hasse had in mind when he wrote "as is easily seen". His own Local-Global Principle, which he conjectured at this colloquium, can be used to prove that his assertion is *not* true over number fields. In any case, we observe that Hasse did not write that Albert's proof "is not valid" but he wrote it "cannot be valid"; this indicates that he had not checked Albert's proof in detail but had in mind some construction of those algebras which would yield a counterexample to Albert's assertion. We do not know when Hasse had discovered that his construction did not work. Maybe this was shortly before he actually went to Göttingen, and then clearly he would not have mentioned it in his talk. Maybe it was after his talk in the discussion with Emmy Noether who, having had her own experience with Albert's paper (as we have seen in the foregoing section), had now studied it in detail and could assure Hasse that it was correct. In any case we know that Hasse, after this experience, now wished to establish contact with Albert in order to clear up the situation.

We do not know Hasse's first letter to Albert, or the precise date when it was sent. Albert's reply is dated February 6, 1931 and so, taking into consideration the time for overseas postal delivery [65] Hasse wrote his letter shortly before or shortly after his Göttingen colloquium lecture on January 13. Thus started the correspondence between Hasse and the 7 years younger Albert, which continued until 1935. There are 15 letters preserved from Albert to Hasse and 2 letters in the other direction.

As we learn from Albert's reply, Hasse had addressed his first letter to Dickson (with whom he had exchanged reprints in the years before) who forwarded it to Albert. At that time Albert was at Columbia University in New York. Hasse had described his own work in his letter; Albert replied that he was very interested in it and he introduced himself. The next letter of Albert is dated March 23, 1931. As we have said earlier, only the letters from Albert to Hasse are preserved while most of the letters from Hasse to Albert seem to be lost. Accordingly, when in the following we report about letters from Albert to Hasse we have to remember that usually between two such letters there was at least one from Hasse to Albert. [66] Now, on March 23 there is already some mathematical discussion in the letter. Replying to Hasse's question on the existence of non-cyclic division algebras of index 4 over a number field, Albert wrote: [67]

> *The question seems to be a number-theoretic one and I see no way to get an algebraic hold on it. It seems to be a hopeless problem to me after more than a year's work on it.*

We observe that this is the same question which Hasse had asked Brauer a year before (letter of March 16, 1930) but could not get an answer either (see p. 54). It appears that the motivation of Hasse was to secure his conjecture concerning the Main Theorem; if the experts were not able to construct non-cyclic algebras over a number field then this would add some point to the conjecture being true.

---

[65] This was about two weeks – there was no air mail yet.

[66] Sometimes there were more than one, for Albert on June 22, 1932 wrote: *"You are, may I say it, a very pleasing friend to write me so often without receiving any answer."*

[67] The letters between Albert and Hasse were written in English. We are citing directly from the original.

Albert also reported on the results of his new paper [Alb:1931a] which had appeared in the January issue of the Transactions of the AMS. Every central division algebra of index 6 [68] over any field of characteristic 0 is cyclic – provided it satisfies what he calls a mild assumption R_2 (and which he could remove in his later paper [Alb:1931c]). In reply to some other related question, namely about the product of two central simple algebras, Albert presented an erroneous answer, saying that $A \otimes_K A$ is a total matrix algebra. But he corrected this four days later, writing that it had to be the product of A with the reciprocal algebra A'. [69]

This and more was of course known already to the trio Brauer-Hasse-Noether, by Brauer's theorems of 1928 which we have cited in Section 4.2 (p. 17), and also by Emmy Noether's Göttingen lectures 1929/30 – which however, were not yet published. [70] In the next letter, dated May 11, 1931 Albert wrote that he had completed a paper containing most of Brauer's results but which he had obtained independently and with his own new methods. Fortunately, he added, he had discovered Brauer's papers before it was too late and hence could give Brauer priority. This refers to Albert's paper [Alb:1931c]. The methods of Albert in this paper are independent of representation theory and, in this sense, they can be regarded as a simplification of Brauer's approach.

We see that Hasse's initiative to open direct contact between German and American mathematicians working on algebras, had from the start been accepted by Albert. We have to be aware of the fact that at the time there were not many international meetings to establish contacts, no e-mail, and journals arrived usually much later than the time when the results had been discovered. The letters of Albert show that he was fully aware of the possibilities which the correspondence with Hasse opened to him: To present his ideas and results to Hasse (and hence to the German group working on algebras) and at the same time to learn about their methods and results (which he regularly shared with Dickson and Wedderburn). In his letter of May 11 he wrote:

> *Your work on quadratic forms is not new to me. In fact I have been reading your Crelle and Jahresbericht work ever since your first letter to me. In this period I have also been able to apply your most fundamental result on quadratic forms in $n \geq 5$ variables, together with my above mentioned new methods to prove the following results...*

It appears that Hasse had explained to Albert his idea of using the Local-Global Principle for quadratic forms, perhaps in a similar way as he had written to Brauer

---

[68] Albert spoke of algebras of "order 36", thereby defining the "order" as the dimension of the algebra as vector space over the base field. We shall avoid this terminology which is in conflict with the terminology used by Noether and Hasse.

[69] Albert corrected this error in the already published paper [Alb:1931a] by putting a page of *Errata* into the same Transactions volume 33.

[70] See [Noe:1929], [Noe:1933], [Noe:1983], [vdW:1931].

(see p. 22). And Albert had reacted immediately, proving theorems about quaternion algebras and algebras of 2-power index which are the obvious candidates whose norm forms may possibly be handled by quadratic forms. Among the results which Albert reports in this letter is his answer to Hasse's former question which he (Albert) had classified as hopeless even in his foregoing letter, namely:

> *Every central division algebra of index* 4 *over an algebraic number field is cyclic.* [71]

Moreover he writes that, over a number field, the product of two quaternion algebras is never a division algebra, and that the same is true for division algebras of 2-power index. Finally, for division algebras of 2-power index he proved Hasse's exponent-index conjecture.

Albert adds that the last of the above statements would probably remain true if 2 is replaced by any prime number p *provided* he could prove some kind of Local-Global Principle for the norm forms of division algebras of index p. But in the next letter (June 30) he apologizes for having given the impression that he could prove the exponent-index conjecture generally. He is still working on it. Also he writes:

> *I want to remark in this connection that I have proved that your results imply that the direct product of any two central division algebras is a division algebra if and only if the indices of the two algebras are relatively prime (for an algebraic reference field).*

And so on. The letters from Albert are full of information about his results, some of them obviously following Hasse's suggestions. Albert writes:

> *The work of the German mathematicians on algebras is very interesting to me and I should like to know all of it if possible... and am very pleased and thankful for the opportunity to communicate with you and know of your results.*

In this connection we have to mention Hasse's "American paper" [Has:1932a]. We have already had several occasions to cite this paper. It contains a comprehensive treatment of cyclic algebras over number fields. The Local-Global Principle for cyclic algebras over number fields is cited there and used in an essential way. We have seen in Section 7.3 (p. 60) that it was around March 6, 1931 when Hasse, as a follow-up to his skew congress, had obtained the Local-Global Principle for

---

[71] One year later, in April 1932 Albert presented to the American Mathematical Society for publication a construction of non-cyclic division algebras of index 4, defined over the rational function field of two variables [Alb:1932a]. We have already mentioned this in the foregoing section; see footnote [58]). These algebras have exponent 2. Some months later, in June 1932, he was able to give a refined construction. this time of non-cyclic division algebras which have both index and exponent 4. Albert was 26 then, and throughout his career he seemed to have been quite proud of this achievement.

cyclic algebras of arbitrary index. On the other hand, Hasse's American paper was received by the editors on May 29, 1931. Thus Hasse had conceived and completed this paper in about two months. In an introductory paragraph to this paper Hasse says:

> *I present this paper for publication to an American journal and in English for the following reason:*

> *The theory of linear algebras has been greatly extended through the work of American mathematicians. Of late, German mathematicians have become active in this theory. In particular, they have succeeded in obtaining some apparently remarkable results by using the theory of algebraic numbers, ideals, and abstract algebra, highly developed in Germany in recent decades. These results do not seem to be as well known in America as they should be on account of their importance. This fact is due, perhaps, to the language difference or to the unavailability of the widely scattered sources...* [72]

Reading this text and knowing that it has been written in the months March to May 1931 when the first letters Hasse-Albert were exchanged, we cannot help feeling that to write this paper in English and to publish it in an American journal, was meant predominantly as a source of information for Hasse's correspondence partner Albert. It seems that Hasse had observed Albert's high qualifications and great power as a mathematician, and he knew that Albert was eager to absorb the methods and results which had been developed in Germany. And so when Hasse wrote the paper, he had in mind Albert as the first and foremost reader. In fact, Albert informed him (letter of November 6, 1931) that he *"was fortunate to read your paper for the editors of the Transactions"*. In other words: Albert had to referee the paper, and so he was the first in America to know its contents, long before the paper finally appeared in 1932. [73]

Between Albert's letter of June 30, 1931 and his next letter of November 6 there is a gap of several months. On Hasse's side, this gap can perhaps be explained because in the summer semester of 1931 he had Harold Davenport as a visitor from England. Hasse had invited Davenport to stay in his Marburg family home under the condition that they would speak English only, so Hasse could

---

[72] The last mentioned fact seems to have been quite serious in those times. Even a big and renowned institution like Columbia University of New York (where Albert stayed in the summer of 1931) did not have the journal "Hamburger Abhandlungen" with the important papers of Artin in its library; for in his letter of May 11, 1931 Albert asks Hasse for information about what is contained in those papers because he was not able to obtain them.

[73] In [Cur:1999] it is said (p. 232) that the results mentioned in Hasse's introductory paragraph had already been well known to Albert. We believe that the Albert-Hasse correspondence of 1931 shows that *the Hasse letters* were what stimulated Albert to study eagerly in more detail the work on algebras which was conducted in Germany.

refresh and upgrade his knowledge of the English language. It is conceivable that Hasse, besides his mathematical activities, was now absorbed in his English studies. [74] Moreover, as is well known, the beginning friendship between Hasse and Davenport induced Hasse to become interested in Davenport's work on the solution number of congruences, which later led to Hasse's proof of the analogue of the Riemann hypothesis for elliptic function fields – thereby starting a series of seminal papers of Hasse on algebraic function fields. We know that after the end of the summer semester 1931 the Hasses together with Davenport went on an extended tour through central Europe (in Davenport's car) which ended only in September at the DMV-meeting in Bad Elster. [75] In view of all this it is plausible that in this summer Hasse did not find the time to work intensively on algebras and to write letters about it.

However in October 1931 Hasse seemed to have again taken up his work on algebras. And he informed Albert about his results. Albert's reply of November 6 starts with the following text:

I received your very interesting communication this morning and was very glad to read of such an important result. I consider it as certainly the most important theorem yet obtained for the problem of determining all central division algebras over an algebraic number field.

What was the result that Hasse had communicated to him, which Albert classified as "the most important result yet obtained..."? Taking into account the delivery time for overseas mail, we conclude that Hasse may have dispatched his letter around October 20. At that time, as we know, Hasse had not yet obtained a proof of the Main Theorem. But we remember from Section 3 (p. 10) that Hasse had already succeeded with the proof that every *abelian* algebra is cyclic, and that he had informed Emmy Noether about it. He had also informed Brauer. Now we see that Hasse had also informed Albert at the same time.

We conclude that by now Hasse had accepted Albert as a correspondence partner on the same level as he had Richard Brauer and Emmy Noether. Thus the "triangle" of Brauer, Hasse and Noether had become a "quadrangle" by the addition of Albert, the latter however being somewhat apart because of the longer distance, which implied a longer time for the transmission of mutual information.

This disadvantage of longer distance became apparent soon.

---

[74] Perhaps Hasse had written to Albert about his English studies, for in the letter of June 30 Albert wrote: *Your English is very clear and understandable. I only wish I could write German half so well!*

[75] For all this see our paper [Roq:2004].

8.3 Albert's Contributions

Albert in his above mentioned letter of November 6 informed Hasse about his results which he had obtained during the summer (when there was no exchange of letters with Hasse), some of which he had already submitted for publication. Together with Hasse's new results on abelian algebras they would lead to interesting consequences, Albert wrote. And he proposed a joint paper with Hasse.

But before this letter reached its destination, Hasse had found the proof of the full Main Theorem. We know from Section 3 that this happened on November 9, when Hasse had received from Brauer and Noether the relevant information. The new proof [76] turned out to be so simple ("trivial" as Noether had called it) that Hasse's former results and methods developed in this direction became obsolete. Thus Albert's letter of November 6 was now superseded by the new development. Nevertheless, Albert had an independent share in the proof of the Main Theorem, as was expressed in the Albert-footnote. Let us describe this in some more detail.

The Albert-footnote consists of three parts. The first part was written when Hasse composed the first draft for the Brauer-Hasse-Noether paper; this was on November 10 as we have seen in Section 3 (p. 12). At this time Hasse had not yet received Albert's letter of November 6, and so he mentioned Albert's contributions which he knew at that time, i.e., those which were contained in the Transactions papers [Alb:1931a] and [Alb:1931c]. These papers had been anounced to him by Albert in his letters of March 23 and May 11 respectively. In particular, "theorem 23" of [Alb:1931c] is mentioned in the Albert-footnote. This theorem reads as follows:

Theorem 23. *Let A be a central division algebra over K of prime power index $p^s > 1$, and M a maximal commutative subfield of A. Then there exists a field extension $L_0|K$ of degree prime to p such that $A_{L_0} = A \otimes_K L_0$ is a central division algebra over L_0 with maximal commutative subfield $L = M \otimes_K L_0$, such that there is a chain of fields*

$$L_0 \subset L_1 \subset \cdots \subset L_{s-1} \subset L_s = L$$

where each $L_i|L_{i-1}$ is cyclic of degree p $(1 \leq i \leq s)$. [77]

Comparing this with the reductions of Brauer and Noether as presented in Section 4.2 (p. 18) we observe that both statements are very similar to each other. Moreover, Albert's proof of "theorem 23" contained the same ingredients as the

---

[76] More precisely: that part of the proof which consisted in the reductions 2 (Brauer) and 3 (Noether).

[77] The notation is ours, not Albert's. – Albert formulates this theorem for base fields K of characteristic 0 only, but from the proof it is clear that it holds for any field of characteristic $\neq p$.

Brauer-Noether "reductions 2 and 3", namely a Sylow argument (like Brauer) and some kind of induction argument from L_0 to L (like Noether). This was the reason why Hasse in the Albert-footnote mentioned "theorem 23" in connection with Brauer and Noether.

But at that time it seemed not yet to be clear whether "theorem 23" indeed was sufficient to *replace completely* the Brauer-Noether arguments. This question was cleared up only later when Albert's letter of November 6 had reached Hasse.

Before discussing this and the later letters of Albert let us report on the reaction of Emmy Noether when she read Hasse's Albert-footnote. We recall from Section 3 that Hasse had sent a draft of their joint paper to Noether, and she commented on it in her replies. In her letter of November 12 she writes concerning the Albert-footnote (i.e., its first part):

> *But you have to weaken your reference to Albert; he has (theorem 19) shown for* cyclic *algebras only, that every prime divisor of the index divides the exponent, and I cannot find anything of the general Brauer reduction.*

Theorem 19 refers to the paper [Alb:1931a] of Albert. It seems that Hasse in his reply protested and pointed out that Albert indeed had essentially the full theorem in question. For in Noether's next letter on November 14, obviously replying to Hasse's "protest", she writes:

> *I have again checked Albert; also in Theorem 20 only cyclic algebras are investigated, and later too he keeps the assumption of cyclicity....*

And she proposes that Hasse should change the footnote; it should be said that Albert did not have the full result, only in the cyclic case. But Hasse seems to have insisted on his point of view, and to have explained the situation to Emmy Noether. For in her next letter of November 22, she wrote:

> *It is good that you have settled the Albert case. Since the fascicles were still unbounded it did not come to my mind to look at the others since I believed I had the paper in question. It now seems to me that he really is very able!* [78]
> *Now I quite agree with your footnote.*

In other words: The volume of the journal in question (Transactions AMS vol. 33) came in several parts (the whole volume had 999 pages!) and those parts were still not bound together in the Göttingen library when Noether looked for Albert's paper. Noether had studied only one of those parts and, hence, read only one of Albert's papers. [79] So she was not aware about all the relevant results of Albert. She had been advised of this by Hasse and now she was happy that Hasse had settled the case in the footnote. And Noether added the remark:

---

[78] *"Er scheint mir also wirklich etwas zu können!"* Certainly, this comment from Emmy Noether means high praise for Albert.

[79] Besides the two papers [Alb:1931a], [Alb:1931c] which are of relevance here, Albert had a third paper in the same volume, namely [Alb:1931b].

By the way, the fact that now all people find this proof, is a consequence of the fact that you have found it first. What was lacking was trivial for everybody who was not completely absorbed in the details of proof as you have been.

Here she refers to the fact that in the case of cyclic algebras Hasse had already proved the Local-Global-Principle in [Has:1931b]. And the generalization to arbitrary algebras she now considered as being "trivial for everybody" (although she and Brauer and Albert had had a hard time to do it). From today's viewpoint we would agree with her. But we have already stated earlier that we should not underestimate the difficulties which former generations of mathematicians had to overcome before they could settle the questions which seem to be trivial for us today.

Comparing the dates: The last mentioned Noether letter had been written on November 22, in reply to a letter of Hasse. Albert had dispatched his letter to Hasse on November 6. So we may assume that Hasse had received Albert's letter around November 20, upon which he had inserted the second part ("Added in proof") into the Albert-footnote, and had Noether informed about this. Noether's letter of November 22 which we have cited above would have been her reply to this.

In this second part of the Albert-footnote Hasse listed three results a), b), c) of Albert, which Albert had mentioned to him in his letter of November 6. However there arises some question concerning statement c). We have checked Albert's letter and found that Hasse's statement is precisely as Albert had written to him. But this statement is *not* what Albert has proved in his papers and what in later letters he referred to. The difference is that in statement c) the degree of Ω' (in Hasse's notation) over the base field is required to be prime to p, whereas Albert later in his letters and in his work [80] does not insist that this is the case. In fact, using the above "theorem 23" it is not difficult to show that the field $\Omega' := L_{s-1}$ has the property that $D \otimes_K L_{s-1}$ is similar to a cyclic division algebra of index p. But the degree $[L_{s-1} : K]$ is divisible by p^{s-1} and not prime to p.

This weak form of c) (where it is not required that the degree of Ω' is prime to p) appears as theorem 13 in Albert's [Alb:1931d], so let us simply call it "theorem 13" in the following, as Albert does in his letters. This "theorem 13" is quite sufficient for the proof of the Local-Global Principle in case K is an algebraic number field, as is easily seen.

For, suppose that the non-trivial division algebra D over K splits everywhere; we may suppose without restriction of generality that the index of A is a prime power $p^s > 1$. According to "theorem 13" there is a finite field extension K' of K such that $D' = D \otimes_K K'$ is of index p and has a cyclic splitting field. Now, since D splits everywhere so does D'. By the Hilbert-Furtwängler Norm Theorem it follows that D' splits, i.e. has index 1. Contradiction.

---

[80] Including Albert's Colloquium Publication "Structure of algebras" [Alb:1939].

We see that indeed, the Local-Global Principle follows "just in a few lines" from Albert's result, as has been said in [Zel:1973]. However, by looking more closely into the matter it turns out that "theorem 13" is based on "theorem 23" whose proof uses the same arguments as Brauer-Noether (as we have said above already). Conversely, "theorem 13" is immediate if one uses the chain of arguments given by Brauer and Noether (p. 18). In other words: *both methods, that of Brauer-Noether and that of Albert, are essentially the same, the differences being non-essential details only.*

It is not an unusual mathematical story: A major result is lurking behind the scenes, ready to be proved, and more than one mathematician succeeds. So this happened here too.

8.4 The Priority Question

Let us return to November 9, 1931, the date when Hasse found the last steps in the proof of the Main Theorem. We have seen in Section 3 that Hasse immediately informed Noether and Brauer about it. But he also informed Albert. Hasse's letter to Albert was dated November 11. This letter crossed paths with Albert's of November 6. Albert received it on November 26, and his reply is dated November 27. There he congratulated Hasse to the "remarkable theorem you have proved". But in the meantime, he added, he had already obtained results which also could be used to prove the Local-Global Principle; they are contained in a paper (which Albert had not mentioned to Hasse before) in the Bulletin of the AMS. [Alb:1931d]. Albert pointed out that it had been submitted on September 9 and the issue of the Bulletin had already been delivered in October. It contained (among other results) the "theorem 13" which we have just discussed. Albert writes:

> As my theorems have already been printed I believe that I may perhaps deserve some priority of your proof. I may say, however, that the remarkable part of your proof for me is the obtaining of the cyclic field. I of course knew your theorem 3.13.

It is not clear what theorem Albert refers to when he cites "3.13". Neither the Brauer-Hasse-Noether paper nor Hasse's American paper contains a theorem with this number. From an earlier part of Albert's letter it seems probable that "theorem 3.13" may stand for the Norm Theorem (see Section 4.1). But in Hasse's American paper this has the number (3.11); thus it may have been just a misprint on the side of Albert.

When Albert speaks of "obtaining of the cyclic field" to be the "remarkable part" of Hasse's proof then he refers to the Existence Theorem which we have discussed in Section 5.2. This shows that Albert immediately saw the weak point of the Brauer-Hasse-Noether paper. For, the existence of the required cyclic field was *not* proved in the Brauer-Hasse-Noether paper. Albert expressed his hope that the existence of the cyclic field

*. . . will be made clear when you publish your proof. In all my work on division
algebras the principal difficulty has been to somehow find a cyclic splitting
field. This your p-adic method accomplishes.*

Probably Albert meant not just *some* cyclic splitting field but a splitting field of
degree equal to the index of the division algebra. As we have seen in Section 5.2
the story of this Existence Theorem is not quite straightforward. But if he really
wanted only to find a cyclic splitting field of unspecified degree, then the proof of
the relevant weak existence theorem was contained in Hasse's paper [Has:1933],
as we had mentioned already in Section 5.4.1.

Hasse responded to Albert's wish for "some priority" by extending the Albert-
footnote a second time, adding a third part where he stated what Albert had
written to him in his letter of November 26. But when Hasse wrote in the
footnote that Albert's paper is "currently printed" (*im Druck befindlich*), then
this may have been a misinterpretation of Albert's text in the letter, which reads
as follows:

*The part of the proof which you attribute to Brauer and Noether is already
in print.*

Obviously, Albert meant that the paper has already appeared (namely in October)
and thus is available in printed form. The German translation of "in print" would
be "*im Druck*". But in German, if it is said that some paper is "*im Druck*" then
the meaning is that the paper is "in press", i.e., in the process of being printed.
This may have led Hasse to the wrong translation of Albert's text. In any case,
neither Hasse nor Noether nor Brauer had yet seen Albert's Bulletin paper. If
indeed it had appeared in October in the U.S.A. then it was not yet available in
German libraries in the beginning of November.

We conclude that Albert's Bulletin paper and the Brauer-Hasse-Noether paper had been written independently of each other. On the other hand, as Albert
points out correctly, his results in his Bulletin paper can be used to prove the
Local-Global Principle for algebras and hence indeed constitute an "independent share" in the proof of the Main Theorem. We have discussed this in the
foregoing section.

Responding to Albert's wish for priority, Hasse did two more things besides
extending the Albert-footnote a second time. First, he sent Albert a copy of the
proof sheets of the Brauer-Hasse-Noether paper, so he could check in particular
the actual text of the Albert-footnote. Secondly, Hasse suggested that they write
a joint paper, to be published in the Transactions, documenting the sequence of
events which led to the proof of the Main Theorem, on the Albert side as well as
on the side of Brauer-Hasse-Noether. Albert should write up the article. In his
letter of January 25, 1932 Albert reported:

*I have finally found time to write up the article by both of us "A determination
of all normal division algebras over an algebraic number field" for the*

Transactions. I gave a historical sketch of the proof, my short proof and a slight revision (to make it more suitable for American readers) of your proof.

This 5-page paper [AlHa:1932] appeared in the same volume as Hasse's American paper [Has:1932a]. It documents for the mathematical community the chain of events which we have extracted here from the Albert-Hasse correspondence.

Irving Kaplansky in his memoir on Albert [Kap:1980] writes:

In the hunt for rational division algebras, Albert had stiff competition. Three top German algebraists (Richard Brauer, Helmut Hasse, and Emmy Noether) were after the same big game... It was an unequal battle, and Albert was nosed out in a photo finish. In a joint paper with Hasse published in 1932 the full history of the matter was set out, and one can see how close Albert came to winning.

This comparison with a competitive sports event reads nicely but after studying the Albert-Hasse correspondence I have the impression that it does not quite reflect the situation. In my opinion, it was not like a competitive game between Albert on the one hand and the trio Brauer-Hasse-Noether on the other. Instead it was teamwork, first among Brauer, Hasse, Noether and then, starting March 1931, Albert joined the team as the fourth member. Within the team, information of any result, whether small or important, was readily exchanged with the aim to reach the envisaged common goal. If a comparison with a sports event is to be given, then perhaps we can look at it as a team of mountaineers who joined to reach the top. The tragedy was that one of the team members (Albert) in the last minute lost contact with the others (because communication was not fast enough) and so they reached the summit on different trails in divided forces, 3 to 1.

Nevertheless, Albert in this situation was upset, which of course is quite understandable. Zelinsky in [Zel:1973] writes that "*Albert was hurt and disappointed by this incident.*" I may be allowed to cite Professor Zelinsky's answer when I asked him about his memories of Albert.

Perhaps my use of the word "hurt" was injudicious, since besides the mental pain that Albert must have felt, the word could connote feelings that he had been taken advantage of, that his correspondence with Hasse was used without due consideration. I have no evidence that he felt that way. You are correct, he was content at last with the resolution of the priority questions. And by the time I knew him, he had become established as a leading mathematician in his own right, which surely affected his attitude toward events of the previous decade.

We believe that, indeed, one can infer from the letters of Albert that he was content with the solution of the priority question as offered by Hasse. The exchange of letters with mathematical results continued in a friendly tone. We have the impression that from now on the tone of the letters was even more open and free. At one time Albert criticised an error which Hasse had made in a previous letter, and at another occasion Hasse pointed out an error of Albert's – in an open and friendly way. At the end of his letter of January 25, 1932 Albert added a somewhat remarkable postscript as follows:

> *Permit me to say I did not believe it possible for mere correspondence to arouse such deep feelings of friendship and comradeship as I now feel for you. I hope you reciprocate.*

We do not know Hasse's answer. In Albert's next letter (April 1, 1932) he thanks Hasse for sending *"photographs and books"*. Moreover we read in this letter:

> *I am very pleased to have been asked to write a report on linear algebras for the Jahresbericht. I shall certainly accept this kind of proposition... I shall study your report and try to understand better precisely what type of report you wish me to write.*

Here, "your report" means Hasse's class field report [Has:1930].

It seems that Hasse wished to revive the long tradition of the *Jahresbericht* of the DMV (German Mathematical Society), to publish reports on recent developments in fields of current interest. [81] However, just at the same time there arose a stiff competition, in the concept of Springer-Verlag to initiate a series of books called *"Ergebnisse der Mathematik"*. We do not know whether it was this competition or there were other reasons that the *Jahresbericht* report series was not continued by the DMV. In fact, Hasse's report [Has:1930] turned out to be the last one in this series and the plan to publish Albert's report in this series was not realized. [82]

In some of the next letters which followed, Albert still tried to convince Hasse that the arguments in the Brauer-Hasse-Noether paper included unnecessary complications whereas his chain of arguments was, in his opinion, shorter and more lucid. This is understandable because as the author he was more familiar

---

[81] One of the best known such reports is Hilberts *"Zahlbericht"* 1897.

[82] As a side remark we may mention that already in 1931 Albert had been asked to write a survey on the theory of algebras, just by the competitor of the *Jahresbericht* series, the newly inaugurated Springer series *"Ergebnisse der Mathematik"*. But after a while this proposal was withdrawn by Otto Neugebauer, the editor of the series. According to Albert (letter of December 9, 1931) Neugebauer wrote that it had been arranged with Deuring to write a survey on "Hypercomplex Systems", and that he (Neugebauer) had discovered just now only that this is the same subject as "Algebras". Deuring's book appeared 1934 with the title "Algebren". Actually, we know that Neugebauer had first approached Emmy Noether to write such a survey but she declined and, instead, recommended her "best student" Deuring for this task. As to Deuring's book, see also [Roq:1989]. – It seems that Albert after these experiences decided to write his book on the Structure of Algebras [Alb:1939] independently and publish it elsewhere.

with his own version. But we have the impression that Hasse was not convinced, although he indeed had high respect for Albert's achievements which had been obtained in a relatively short time after their correspondence started. Sometimes Hasse scribbled some marginal notes on Albert's letters as reminders for his reply, and from this we can extract at least some marginal information about Hasse's letters to Albert. On April 1, 1932 Albert wrote that he believed

> ... *your whole Transactions paper could be simplified considerably if this reduction had been made to begin with. Of course it is a matter of personal taste and you may even yet not agree.*

Here, Albert has in mind the reduction from arbitrary algebras to those of prime power index by means of Brauer's product theorem which he, Albert, had discovered independently.

To the second sentence we find the marginal note "*yes!*" from Hasse's hand on the letter, which seems to imply that his personal taste was somewhat different. After all, as we have seen in Section 4.2, Hasse's first draft contained this reduction to prime power index, and it was Emmy Noether who threw this away because it was unnecessary. – The first sentence is commented by "*no! (class field theory!)*". Here Hasse refers to the close connection of the theory of algebras to class field theory – something which was outside of the realm of interest of Albert, except that he admitted class field theory as a means to prove theorems on algebras, if necessary. See also Section 6.3.2 .

8.5 Remarks

It seems that Hasse had invited Albert to visit him in Germany. In his letter of June 30, 1931 Albert wrote:

> *I hope that in perhaps two years I may visit Germany and there see you and discuss our beautiful subject, linear algebras.*

On November 27, 1931 the plans had become more specific:

> *I am very glad that you are interested in the possibility of my visiting you. I hope that I will be able to leave Chicago on Sept 1, 1933 to return here not later than Dec 31, 1933. I do not believe I can make the trip before that time.*

Due to the disastrous political events which took place in Germany in the year 1933 these plans could not be realized. Instead, Albert applied for and received an appointment at the Institute for Advanced Study in Princeton for the academic year 1933/34. There are two letters from Albert in Princeton to Hasse which are preserved. They do contain interesting material but this is not immediately connected with the Main Theorem, hence we will not discuss it here.

While Albert was in Princeton in 1933/34 he met two-thirds of the German team, Richard Brauer and Emmy Noether, who had been forced to leave their country. On the relationship between Albert and Brauer, Mrs. Nancy Albert, the daughter of A. A. Albert, reports: [83]

> *I have news about Brauer from letters written to my parents. When Brauer arrived in America from Germany, he spoke little English, and was rather traumatized from all that he had been through. He stopped at Princeton in 1933. My father took Brauer under his wing, made him feel welcome, and took him to meet Wedderburn. Later, my father put together a large mathematical conference in Chicago, where the Alberts hosted a large dinner party, and the Brauers became good friends of my parents. Their relationship continued the remainder of my father's life.*

Mathematically, however, Albert's and Brauer's work went in somewhat different directions. Albert continued to work on algebras, including more and more non-associative structures. Brauer concentrated on group theory and representations. Most of the work on finite simple groups and their classification can be traced to his pioneering achievements, and he advanced to *"one of the leading figures on the international mathematical scene"* (J. A. Green).

About the Princeton contact of Albert with Noether we have the following information. In an undated letter to Hasse from Princeton, probably written in January 1934, Albert writes:

> *I have seen R. Brauer and E. Noether. They passed through here and stayed a short while.*

And on February 6, 1934:

> *E. N. speaks here tomorrow on Hypercomplex numbers and Number theory.*

Emmy Noether herself, in her letters to Hasse, is somewhat more detailed about the mathematical life in Princeton. In a letter of March 6, 1934 she writes the following report, and we note that Albert is mentioned:

> *. . . I have, since February, started a lecture in Princeton once a week – at the Institute and not at the "men's"-university which does not admit anything female. . . At the beginning I have started with representation modules, groups with operators. This winter Princeton is treated algebraically, for the first time but quite thoroughly. Weyl also lectures about representation theory but will soon switch to continuous groups. Albert, in a "leave of absence" there, has last year lectured on something hypercomplex in the style of Dickson, together with his "Riemann matrices". Vandiver, also "leave of absence", lectures on number theory, the first time in Princeton since time immemorial. And after I had given a survey on class field theory in the Mathematics Club, von Neumann has ordered twelve copies of Chevalley as a textbook (Bryn Mawr also shall get a copy). On this occasion I was told that your Lecture*

---

[83] Personal communication.

Notes will be translated into English, now hopefully in sufficiently many copies – I had recommended this already in the fall. My audience consists essentially of research fellows, besides Albert and Vandiver, but I noticed that I have to be careful; these people are used to explicit computations, and some of them I have already driven away!

We can safely assume that Albert was not one of the dropouts from Noether's course. He knew about the importance of Emmy Noether's viewpoint on algebra and on the whole of mathematics. Noether's ideas have often been described and so we will not repeat this here. [84] But at the time we are considering, Noether's ideas had not yet penetrated mathematics everywhere. Albert himself had his training with Dickson, and his papers in those first years of his mathematical activity were definitely "Dickson style". It was only gradually that Albert started to use in his papers the "Modern Algebra" concepts in the sense of Emmy Noether and van der Waerden. In 1937 he published the book "Modern higher algebra" [Alb:1937] which was a student textbook in the "*modern*" (at that time) way of mathematical thinking. [85]

Albert explicitly stated that his textbook was meant as an introduction to the methods which will be used in his forthcoming book on algebras. That second book appeared in 1939 [Alb:1939] with the title "Structure of Algebras", and it was also written with the viewpoint of "Modern Algebra". [86] It seems to us that to a large degree this was a direct consequence of Albert being exposed at Princeton to Emmy Noether's influence.

In the preface to his book Albert says:

The theory of linear associative algebras probably reached its zenith when the solution was found for the problem of determining all rational division algebras. Since that time it has been my hope that I might develop a reasonably self-contained exposition of that solution as well as of the theory of algebras upon which it depends and which contains the major portion of my own discoveries.

We do not intend here to give a review of Albert's book which, after all, is well known and has become a classic. It is our aim here to point out that to a large extent the book is the outcome of his participation in the team together with Hasse, Brauer and Noether – notwithstanding the fact that the book contains also other aspects of the theory of algebras, e.g., Riemann matrices and p-algebras.

---

[84] See, e.g., Hermann Weyl's obituary address in Bryn Mawr 1935 [Wey:1935].

[85] The book was refereed in the *Zentralblatt* by Helmut Hasse.

[86] This book was refereed in the *Zentralblatt* by W. Franz, a former Ph. D. student of Hasse.

Epilogue: Käte Hey

In the history of mathematics we can observe not infrequently, that after an important result has been found, it is discovered that the very same result, in more or less explicit form, had been discovered earlier already. This happened also to the Local-Global Principle for algebras which is the basis for Hasse's proof of the Main Theorem.

On January 26, 1933, one year after the appearance of the Brauer-Hasse-Noether paper, the editor of the *Hamburger Abhandlungen* received a manuscript of a paper [Zor:1933] which begins as follows:

> *The theory of the ζ-function of a skew field has been developed in detail by Miss K. Hey in her dissertation (Hamburg 1929). In the present note I would like to draw attention to the arithmetic consequences which are derived there, so that after some correction and streamlining they are recognized as a*
>
> **new proof of a main theorem on algebras and of the general quadratic reciprocity law.**
>
> *The said main theorem on algebras is the basis for deriving the reciprocity law with non-commutative methods; therefore its independent foundation is important for methodical reasons.*

The "main theorem" which is meant here is not quite the Main Theorem of Brauer-Hasse-Noether but the Local-Global Principle as formulated in Section 4. The "general quadratic reciprocity law" is extra mentioned by the author because it follows directly from Käte Hey's treatment in the case of quaternion algebras. In the next sentence however, "reciprocity law" means Artin's reciprocity law; to derive this from the Main Theorem one had to follow Hasse's method as explained in Section 6.

The author of this article was Max Zorn, a former Ph. D. student of Artin in Hamburg. [87] He had been the second Ph. D. student of Artin, the first one had been Käte Hey whose thesis he is referring to in his note. She received her degree

---

[87] Zorn (1906–1993) received his Ph. D. 1930 with a paper on alternative algebras. In 1933 he was forced by the Nazi regime to leave Germany . His name is known to the mathematical community through his "Zorn's Lemma".

in 1927. [88] Her thesis [Hey:1929] had never been published in a mathematical journal but it was printed, and was distributed among interested mathematicians. We know that Hasse and Emmy Noether each owned a copy, perhaps Richard Brauer too. The thesis was refereed in the *Jahrbuch für die Fortschritte der Mathematik*, vol. 56.

The aim of Hey's thesis was to extend the known methods of analytic number theory, in particular those of Hecke leading to the functional equation of zeta functions of number fields, to division algebras instead of number fields. She defined the Zeta function $\zeta_D(s)$ of a division algebra D whose center K is an algebraic number field. But she considered only the finite primes $\mathfrak{p}$ of K. If that function is supplemented by factors corresponding to the infinite primes of K (which today is the standard procedure) then the analytic treatment of that extended function, including its functional equation, shows that, if compared with the zeta function $\zeta_K(s)$ of the center, it admits two poles (if $D \neq K$), which in some way correspond to primes $\mathfrak{p}$ which are ramified in D, i.e., at which $D_{\mathfrak{p}}$ does not split. Indeed, the existence of such primes is the content of the Local-Global Principle. Hey used Artin's paper [Art:1928c] on the arithmetic of algebras, and also the full arsenal of analytic number theory known at that time which centered around Hecke's work. Deuring says in [Deu:1934]:

Hey's proof of the Main Theorem represents the strongest concentration of analytic tools to reach the aim.

For a discussion of Hey's thesis and Zorn's note we refer to the recent essay [Lor:2004] by Falko Lorenz.

Hey's thesis is considered to be very difficult to read. It seemed to be generally known at the time, at least among the specialists, that Hey's thesis contained errors. But Zorn points out how those errors could be corrected in a quite natural and straightforward manner.

Thus if Hasse (or Noether, or Brauer, or Albert) had known this earlier, then the proof of the Main Theorem could have been completed earlier. It is curious that Hey's thesis had not been mentioned in the correspondence of Hasse, not with Artin, not with Noether, Brauer or Albert. At least not before Zorn's note became known.[89]

---

[88] Käte Hey (1904–1990) left the university some time after she had obtained her degree, then she became a teacher at a gymnasium. More biographical details can be found in [Lor:2004] and [Tob:1997].

[89] Emmy Noether got to know Zorn's note some time in winter 1932/33. She was so impressed that she suggested to two of her Ph. D. students to continue work in that direction. One of those students was Ernst Witt who in his thesis transferred Hey's results to the function field case. The other student was Wolfgang Wichmann who presented a much simplified proof of Hey's functional equation of the zeta function of a division algebra, however up to a $\pm$ sign only.

Later, when Hasse and Noether discussed how much analysis should and could be used in class field theory, Hasse wrote to her (letter of November 19, 1934):

> *If analysis is to be used in the foundation of class field theory ... then one should aim with Hey's cannon at the norm theorem, the sum formula for the invariants of algebras, and the theorem on algebras splitting everywhere. From there one can aim backwards to class field theory in the classical sense, like on your 50th birthday.*

Here, the reference to Noether's 50th birthday is to be read as the reference to Hasse's paper [Has:1933] which he had dedicated to her on the occasion of that birthday; we have mentioned it several times in this paper.

In his letter Hasse pointed out that no one at that time had been able to develop class field theory without using methods from analytic number theory. His own approach in the "birthday paper" [Has:1933] is based on the Hilbert-Furtwängler Norm Theorem which in turn was proved using analytic methods of zeta functions of number fields. But soon after, Chevalley [Che:1935] succeeded to give a foundation of class field theory free from analysis; see also [Che:1940].

References

[Alb:1930] A. A. Albert, *New results in the theory of normal division algebras.* Trans. Amer. Math. Soc. 32 (1930) 171–195   5, 6, 56

[Alb:1930a] A. A. Albert, *On direct products, cyclic division algebras, and pure Riemann matrices.* Proc. Nat. Acad. Sci. U.S.A. 16 (1930) 313–315

[Alb:1931a] A. A. Albert, *On direct products, cyclic division algebras, and pure Riemann matrices.* Trans. Amer. Math. Soc. 33 (1931) 219–234, Correction p. 999. [Remark: This paper has not been included into the "Collected Papers" of A. A. Albert.]   61, 64, 68, 69

[Alb:1931b] A. A. Albert, *On normal division algebras of type R in thirty-six units.* Trans. Amer. Math. Soc. 33 (1931) 235–243   69

[Alb:1931c] A. A. Albert, *On direct products.* Trans. Amer. Math. Soc. 33 (1931) 690–711   61, 64, 68, 69

[Alb:1931d] A. A. Albert, *Division algebras over algebraic fields.* Bull. Amer. Math. Soc. 37 (1931) 777–784   61, 70, 71

[Alb:1932a] A. A. Albert, *A construction of non-cyclic normal division algebras.* Bull. Amer. Math. Soc. 38 (1932) 449–456   55, 65

[Alb:1932b] A. A. Albert, *On the construction of cyclic algebras with a given exponent.* Amer. J. of Math. 54 (1932) 1–13   34

[Alb:1934] A. A. Albert, *On normal Kummer fields over a non-modular field.* Trans. Amer. Math. Soc. 36 (1934) 885–892   33

[Alb:1936] A. A. Albert, *Normal division algebras of degree p^e over a field F of characteristic p.* Trans. Amer. Math. Soc. 39 (1936) 183–188

[Alb:1937] A. A. Albert, *Modern higher algebra.* Chicago (1937)   77

[Alb:1938] A. A. Albert, *Non-cylic algebras with pure maximal subfields.* Bull. Amer. Math. Soc. 44 (1938) 576–579   34

[Alb:1939] A. A. Albert, *Structure of Algebras.* Amer. Math. Soc. Colloquium Publications, vol. 26 (1939)   70, 74, 77

[AlHa:1932] A. A. Albert, H. Hasse, *A determination of all normal division algebras over an algebraic number field.* Trans. Amer. Math. Soc. 34 (1932) 171–214   27, 28, 73

[Ami:1972] S. A. Amitsur, *On central division algebras.* Israel J. of Math. 12 (1972) 408–420   55

[Art:1927] E. Artin, *Beweis des allgemeinen Reziprozitätsgesetzes.* Abh. Math. Sem. Hamburg 5 (1927) 353–363   42, 45

[Art:1928a] E. Artin, *Über einen Satz von Herrn H. J. Maclagan Wedderburn.* Abh. Math. Seminar Hamburg. Univers. 5 (1928) 245–250   46

[Art:1928b] E. Artin, *Zur Theorie der hyperkomplexen Zahlen.* Abh. Math. Seminar Hamburg. Univers. 5 (1928) 251–260   46

[Art:1928c] E. Artin, *Zur Arithmetik hyperkomplexer Zahlen.* Abh. Math. Seminar Hamburg. Univers. 5 (1928) 261–289   46, 80

[Art:1950] E. Artin, *The influence of J. H. M. Wedderburn on the development of modern algebra.* Bull. Amer. Math. Soc. 56 (1950) 65–72

[ArNeTh:1944] E. Artin, C. J. Nesbitt, R. M. Thrall, *Rings with minimum condition.* University of Michigan Press. X (1944) 123 pp.

[ArTa:1968] E. Artin, J. Tate, *Class field theory.* Benjamin, New York-Amsterdam (1968) 259 pp.   35, 50

[Bran:1930] H. Brandt, *Zur Idealtheorie Dedekindscher Algebren.* Comment. Math. Helv. 2 (1930) 13–17   47

[Bra:1926] R. Brauer, *Über Zusammenhänge zwischen den arithmetischen und invariantentheoretischen Eigenschaften von Gruppen linearer Substitutionen.* Sitzungsberichte Akad. Berlin 1926, 410–416   10

[Bra:1928] R. Brauer, *Untersuchungen über die arithmetischen Eigenschaften von Gruppen linearer Substitutionen. Erste Mitteilung.* Math. Zeitschr. 28 (1928) 677–696   10, 16, 51, 61

[Bra:1929a] R. Brauer, *Über Systeme hyperkomplexer Größen.* Jber. Deutsch. Math. Verein. 38 (1929) 2. Abteilung *47–48* (kursiv)   17

[Bra:1929b] R. Brauer, *Über Systeme hyperkomplexer Zahlen.* Math. Zeitschr. 30 (1929) 79–107   9, 17, 18

[Bra:1930] R. Brauer, *Untersuchungen über die arithmetischen Eigenschaften von Gruppen linearer Substitutionen. Zweite Mitteilung.* Math. Zeitschr. 31 (1930) 733–747   55

[Bra:1932a] R. Brauer, *Über die algebraische Struktur von Schiefkörpern.* J. Reine Angew. Math. 166 (1932) 241–252

[Bra:1932b] R. Brauer, *Über die Konstruktion der Schiefkörper, die von endlichem Rang über ihr Zentrum sind.* J. Reine Angew. Math. 168 (1932) 44–64

[Bra:1933] R. Brauer, *Über den Index und den Exponenten von Divisionsalgebren.* Tôhoku Math. J. 37 (1933) 77–87   34

[Bra:1945] R. Brauer, *On the representation of a group of order g in the field of the g-th roots of unity.* Amer. J. of Math. 67 (1945) 461–471   32

[Bra:1947] R. Brauer, *Applications of induced characters.* Amer. J. of Math. 69 (1947) 709–716   32

[BrHaNo:1932] R. Brauer, H. Hasse, E. Noether, *Beweis eines Hauptsatzes in der Theorie der Algebren.* J. Reine Angew. Math. 167 (1932) 399–404 5, 13, 17, 25, 26, 27, 32, 39, 42, 44, 50, 61

[BrNo:1927] R. Brauer, E. Noether, *Über minimale Zerfällungskörper irreduzibler Darstellungen.* Sitzungsberichte Akad. Berlin 1927, 221–228 16, 54

[BrTa:1955] R. Brauer, J. Tate, *On the characters of finite groups.* Annals of Math. 62 (1955) 1–7 32

[Bru:1997] E. S. Brussel, *Wang counterexamples lead to noncrossed products.* Proc. of the Amer. Math. Soc. 125 (1997) 2199–2206 55

[Che:1933] C. Chevalley *Le théorie du symbole de restes normiques.* J. Reine Angew. Math. 169 (1933) 140–157 40, 49

[Che:1935] C. Chevalley *Sur la théorie du corps de classes.* C. R. Acad. Sci. Paris 201 (1935) 632–634 81

[Che:1940] C. Chevalley *La théorie du corps de classes.* Annals of Math. (2) 41 (1940) 394–418 81

[Cur:1999] C. W. Curtis, *Pioneers of representation theory: Frobenius, Burnside, Schur and Brauer.* History of Mathematics, vol. 15. Published by Amer. Mathematical Society and London Mathematical Society (1999) 287 pp. 51, 52, 57, 66

[Deu:1934] M. Deuring, *Algebren.* Ergebnisse der Mathematik und ihrer Grenzgebiete, Berlin (1934) 21, 80

[Dic:1923] L. E. Dickson, *Algebras and their Arithmetics.* Chicago (1923). Reprint 1938 46

[Dic:1927] L. E. Dickson, *Algebren und ihre Zahlentheorie. Mit einem Kapitel über Idealtheorie von A. Speiser.* Übersetzt aus dem Amerikanischen von J. J. Burckhardt und E. Schubarth. Orell Füssli Verlag Zürich. (1927) 6, 46

[Fei:1979] W. Feit, *Richard D. Brauer.* Bull. Amer. Math. Soc., New Ser. 1 (1979) 1–20 57

[Fre:2003] G. Frei, *Johann Jakob Burckhardt zum 100. Geburtstag am 13. Juli 2003.* Elemente der Mathematik 58 (2003) 134–140 46

[Fur:1902] Ph. Furtwängler, *Über das Reziprozitätsgesetz der ℓ-ten Potenzreste in algebraischen Zahlkörpern, wenn ℓ eine ungerade Primzahl bedeutet.* Abhandlungen der Gesellschaft der Wissenschaften in Göttingen, Mathematisch-Physikalische Klasse, Neue Folge 2 (1902) 16

[Gru:1932] W. Grunwald, *Charakterisierung des Normenrestsymbols durch die p-Stetigkeit.* Math. Annalen 107 (1932) 145–164 29, 30

[Gru:1933] W. Grunwald, *Ein allgemeines Existenztheorem für algebraische Zahlkörper.* J. Reine Angew. Math. 169 (1933) 103–107 29, 30

[Has:1923] H. Hasse, *Über die Darstellbarkeit von Zahlen durch quadratische Formen im Körper der rationalen Zahlen.* J. Reine Angew. Math. 152 (1923) 129–148 53

[Has:1924a] H. Hasse, *Darstellbarkeit von Zahlen durch quadratische Formen in einem beliebigen algebraischen Zahlkörper.* J. Reine Angew. Math. 153 (1924) 113–130 22

[Has:1924b] H. Hasse, *Äquivalenz quadratischer Formen in einem beliebigen algebraischen Zahlkörper.* J. Reine Angew. Math. 153 (1924) 153–162 22

[Has:1926] H. Hasse, *Bericht über neuere Untersuchungen und Probleme aus der Theorie der algebraischen Zahlkörper. I. Klassenkörpertheorie.* Jber. Deutsch. Math. Verein. 35 (1926) 1–55 48

[Has:1926a] H. Hasse, *Zwei Existenztheoreme über algebraische Zahlkörper.* Mathematische Annalen 95 (1926) 229–238 27, 35

[Has:1926b] H. Hasse, *Ein weiteres Existenztheorem in der Theorie der algebraischen Zahlkörper.* Mathematische Zeitschrift 24 (1926) 149–160 27

[Has:1927] H. Hasse, *Bericht über neuere Untersuchungen und Probleme aus der Theorie der algebraischen Zahlkörper. Ia. Beweise zu Teil I.* Jber. Deutsch. Math. Verein. 36 (1927) 233–311 58

[Has:1927a] H. Hasse, *Existenz gewisser algebraischer Zahlkörper.* Sitzungsberichte Akad. Berlin 1927 (1927) 229–234 54

[Has:1927b] H. Hasse, *Über das Reziprozitätsgesetz der m-ten Potenzreste.* J. Reine Angew. Math. 158 (1927) 228–259 43

[Has:1928] H. Hasse, *Besprechung von L. E. Dickson, Algebren und ihre Zahlentheorie* Jber. Deutsch. Math. Verein. 37, 2. Abteilung (1928) 90–97 46

[Has:1930] H. Hasse, *Bericht über neuere Untersuchungen und Probleme aus der Theorie der algebraischen Zahlkörper. II. Reziprozitätsgesetze.* Jber. Deutsch. Math. Verein. Ergänzungsband 6 (1930) 1–204 16, 19, 58, 74

[Has:1930a] H. Hasse, *Neue Begründung und Verallgemeinerung der Theorie des Normenrestsymbols.* J. Reine Angew. Math. 162 (1930) 134–144 43

[Has:1930b] H. Hasse, *Die Normenresttheorie relativ-abelscher Zahlkörper als Klassenkörpertheorie im Kleinen.* J. Reine Angew. Math. 162 (1930) 145–154 43

[Has:1931] H. Hasse, *Über ℘-adische Schiefkörper und ihre Bedeutung für die Arithmetik hyperkomplexer Zahlensysteme.* Math. Annalen 104 (1931) 495–534 25, 26, 34, 37, 38, 39, 44, 45, 46, 47, 54, 57, 59

[Has:1931a] H. Hasse, *Beweis eines Satzes und Widerlegung einer Vermutung über das allgemeine Normenrestsymbol.* Nachr. Ges. Wiss. Göttingen, Math.-Phys. Kl. Math.-Phys. Kl. Fachgr. I Math. Nr. 20 (1931) 64–69 16, 19, 47, 60

[Has:1931b] H. Hasse, *Theorie der zyklischen Algebren über einem algebraischen Zahlkörper.* Nachr. Ges. Wiss. Göttingen, Math.-Phys. Kl. Fachgr. I Math. Nr. 21 (1931) 70–79 60, 70

[Has:1932] H. Hasse, *Ansprache zum 70. Geburtstag des Geh. Regierungsrats Prof. Dr. Kurt Hensel am 29. Dezember 1931.* Mitteilungen Universitätsbund Marburg 1932 – Heft 1. 9

[Has:1932a] H. Hasse, *Theory of cyclic algebras over an algebraic number field.* Trans. Amer. Math. Soc. 34 (1932) 171–214 21, 27, 39, 43, 44, 60, 65, 73

[Has:1932b] H. Hasse, *Additional note to the author's "Theory of cyclic algebras over an algebraic number field".* Trans. Amer. Math. Soc. 34 (1932) 727–730

[Has:1932c] H. Hasse, *Strukturtheorie der halbeinfachen Algebren über algebraischen Zahlkörpern.* Verhandlungen Kongreß Zürich 1932, 2 (1932) 18–19 39, 48

[Has:1933] H. Hasse, *Die Struktur der R. Brauerschen Algebrenklassengruppe über einem algebraischen Zahlkörper. Insbesondere Begründung des Normenrestsymbols und die Herleitung des Reziprozitätsgesetzes mit nichtkommutativen Hilfsmitteln..* Math. Annalen 107 (1933) 731–760 6, 13, 18, 26, 28, 29, 31, 35, 42, 44, 45, 49, 72, 81

[Has:1933a] H. Hasse, *Vorlesungen über Klassenkörpertheorie.* Preprint, Marburg (1933). Later published in book form by Physica Verlag Würzburg (1967) 19, 48

[Has:1933b] H. Hasse, *Berichtigungen zu H. Hasse, Bericht über neuere Untersuchungen und Probleme aus der Theorie der algebraischen Zahlkörper, Teil Ia und Teil II.* Jber. Deutsch. Math. Verein. 42 (1933) 85–86

[Has:1949] H. Hasse, *Zur Frage des Zerfällungskörpers einer endlichen Gruppe.* Math. Nachrichten 3 (1949) 4–6 32

[Has:1950] H. Hasse, *Zum Existenzsatz von Grunwald in der Klassenkörpertheorie.* J. Reine Angew. Math. 188 (1950) 40–64 30, 35

[Hen:1935] K. Henke, *Zur arithmetischen Idealtheorie hyperkomplexer Zahlen.* Abh. Math. Seminar Hamburg. Univers. 11 (1935) 311–332

[Her:1930] J. Herbrand, *Sur les unités d'un corps algébrique.* Comptes Rendus 192 (1931) 24-27. Errata C. R. 192 (1931) 188 59

[Hey:1929] K. Hey, *Analytische Zahlentheorie in Systemen hyperkomplexer Zahlen.* Dissertation Hamburg (1929) 49 pp. 80

[Hil:1897] D. Hilbert, *Die Theorie der algebraischen Zahlkörper.* Jahresbericht der Deutschen Mathematiker Vereinigung 4 (1897) 175–546 16

[Hop:1939] C. Hopkins, *Rings with minimum condition for left ideals.* Annals of Math. II. ser. 40 (1939) 712–730

[Iya:1998] S. Iyanaga, *Travaux de Claude Chevalley sur la théorie du corps de classes. Introduction.* In: Claude Chevalley, Collected works, vol. I. (To appear soon)

[Kap:1980] I. Kaplansky, *Abraham Adrian Albert. November 9, 1905 – June 6, 1972.* Biographical Memoirs, National Academy of Sciences 51 (1980) 3–22   73

[Koe:1933] G. Köthe, *Erweiterung des Zentrums einfacher Allgebren.* Mathematische Annalen 107 (1933) 761–766   26

[Kue:1913] J. Kürschák, *Über Limesbildung und allgemeine Körpertheorie.* J. f. d. reine u. angewandte Math. 142 (1913) 211–253

[Lor:1998] F. Lorenz, *Ein Scholion zum Satz 90 von Hilbert.* Abh. Math. Seminar Hamburg. Univers. 68 (1998) 347–362   17

[Lor:2004] F. Lorenz, *Käte Hey und der 'Hauptsatz der Algebrentheorie'.* Manuskript (2003) 14 pp. (to appear in the *Mitteilungen der Mathematischen Gesellschaft Hamburg*)   5, 80

[LoRo:2003] F. Lorenz, P. Roquette, *The theorem of Grunwald-Wang in the setting of valuation theory.* In: Fields Institute Communications Series, vol. 35 (2003) 175–212   35

[Noe:1925] E. Noether, *Gruppencharaktere und Idealtheorie.* Jber. Deutsch. Math. Verein. 6, 2. Abt. (1930) p. 144 (kursiv)   52

[Noe:1927] E. Noether, *Abstrakter Aufbau der Idealheorie in algebraischen Zahl- und Funktionenkörpern.* Math. Annalen 96 (1927) 26–61

[Noe:1929] E. Noether, *Hyperkomplexe Größen und Darstellungstheorie.* Math. Zeitschr. 30 (1929) 641–692   17, 19, 52, 64

[Noe:1932] E. Noether, *Hyperkomplexe Systeme in ihren Beziehungen zur kommutativen Algebra und Zahlentheorie.* Verhandl. Intern. Math. Kongreß Zürich 1 (1932) 189–194   49

[Noe:1933] E. Noether, *Nichtkommutative Algebra.* Math. Zeitschr. 37 (1933) 514–541   52, 64

[Noe:1983] E. Noether, *Algebra der hyperkomplexen Größen. Vorlesung Wintersemester 1929/30, ausgearbeitet von M. Deuring.* Publiziert in: Emmy Noether, Collected Papers. Springer (1983) 711–763   64

[Roq:1952] P. Roquette, *Arithmetische Untersuchung des Charakterringes einer endlichen Gruppe.* J. Reine Angew. Math. 190 (1952) 148–168   32

[Roq:1989] P. Roquette, *Über die algebraisch-zahlentheoretischen Arbeiten von Max Deuring.* Jber. Deutsch. Math. Verein. 91 (1989) 109–125   74

[Roq:2001] P. Roquette, *Class field theory in characteristic p, its origin and development.* In: Miyake et al. (ed.), *Class field theory – its centenary and prospect.* Advanced Studies in Pure Mathematics 30 (2001) 1–83   52

[Roq:2004] P. Roquette, *The Riemann hypothesis in characteristic p, its origin and development. Part 2. The first steps by Davenport and Hasse.* To appear in: Mitteilungen der Mathematischen Gesellschaft in Hamburg (2004)   67

[Sal:1982] D. Saltman, *Generic Galois extensions and problems in field theory.* Advances in Math. 43 (1982) 250–483

[Schu:1906] I. Schur, *Arithmetische Untersuchungen über endliche Gruppen linearer Substitutionen.* Berl. Akad. Ber. 1906, 164–184 32

[Schu:1919] I. Schur, *Einige Bemerkungen zu der vorstehenden Arbeit von Herrn A. Speiser* Mathematische Zeitschrift 5 (1919) 7–10 17

[Sco:1921] G. Scorza, *Corpi numerici ed algebre.* Messina: Principato (1921). IX, 462 S.

[Spe:1926] A. Speiser, *Allgemeine Zahlentheorie.* Vierteljahrsschrift der Naturforschenden Gesellschaft in Zürich 71 (1926) 8–48 46

[Tau:1979] O. Taussky-Todd, *My personal recollections of Emmy Noether.* In: Brewer, James W. et al. (ed.), Emmy Noether. A tribute to her life and work. New York (1981) 79–92 56

[Tob:1997] R. Tobies, *Die Hamburger Doktorin der Mathematik Käte Hey (1904–1990).* In: *Promotionen von Frauen in Mathematik – ausgewählte Aspekte einer historiographischen Untersuchung.* Mitteilungen d. Mathematischen Gesellschaft Hamburg 16 (1997) 39–63 80

[Tob:2003] R. Tobies, *Briefe Emmy Noethers an P. S. Alexandroff.* N. T. M. 11 (2003) 100–115 58, 59

[vdW:1931] B. L. van der Waerden, *Moderne Algebra*, vol. II (1931) 64

[vdW:1934] B. L. van der Waerden, *Elementarer Beweis eines zahlentheoretischen Existenztheorems.* J. Reine Angew. Math. 171 (1934) 1–3 31

[vdW:1975] B. L. van der Waerden, *On the sources of my book Moderne Algebra.* Historia Mathematica 2 (1975) 31–40

[Wad:2002] A. W. Wadsworth, *Valuation theory on finite dimensional algebras.* In: Valuation Theory and its Applications, vol. I. Ed. by F.-V. Kuhlmann et al. Fields Institute Communications Series, vol. 32 (2002) 385–449 38

[Wan:1948] Sh. Wang, *A counter-example to Grunwald's theorem.* Annals of Math. 49 (1948) 1008–1009

[Wan:1950] Sh. Wang, *On Grunwald's theorem.* Annals of Math. 51 (1950) 471–484 30, 35

[Wey:1935] H. Weyl, *Emmy Noether.* Memorial address delivered in Goodhart hall, Bryn Mawr College, on April 26, 1935. Published in Scripta math. 3 (1935) 201–220 77

[Wha:1942] G. Whaples, *Non-analytic class field theory and Gruenwald's theorem.* Duke Math. J. 9 (1942) 455-473 29

[Wit:1931] E. Witt, *Über die Kommutativität endlicher Schiefkörper.* Abh. Math. Seminar Hamburg. Univers. 8 (1931) 413–413

[Wit:1934] E. Witt, *Riemann-Rochscher Satz und ζ-Funktion im Hyperkomplexen.* Mathematische Annalen 110 (1934) 12–28

[Wit:1983] E. Witt, *Vorstellungsbericht.* Jahrbuch der Akademie der Wiss. Göttingen (1983) 100–101

[Zel:1973] D. Zelinsky, *A. A. Albert.* Amer. Math. Monthly 80 (1973) 661–665 71, 73

[Zor:1933] M. Zorn, *Note zur analytischen Zahlentheorie.* Abh. Math. Seminar
    Hamburg. Univers. 9 (1933) 197–201   79

Added in proof:

G. Frei, *On the history of algebras and their relation to number theory from
    Hamilton to Hasse (1843–1932).* To appear in the Proceedings of the MSRI
    workshop (April 2003) on the history of algebras, eds. Karen Parshall and
    Jeremy Gray.

Index

Printing and Binding: Strauss GmbH, Mörlenbach